Air Force Installation Energy Assurance

An Assessment Framework

Anu Narayanan, Debra Knopman, James D. Powers, Bryan Boling, Benjamin M. Miller, Patrick Mills, Kristin Van Abel, Katherine Anania, Blake Cignarella, Connor P. Jackson

Prepared for the United States Air Force
Approved for public release; distribution unlimited

For more information on this publication, visit www.rand.org/t/RR2066

Library of Congress Cataloging-in-Publication Data is available for this publication.
ISBN: 978-0-8330-9864-1

Published by the RAND Corporation, Santa Monica, Calif.

Preface

New and evolving U.S. Air Force missions are making mission success more and more dependent on installations in the continental United States and assured access to energy. Especially, mission-essential facilities and installations increasingly require access to electric power to provide mission assurance. In recent years, the Air Force has recognized these issues and started to increase its focus on what has variously been referred to as *energy security*, *energy assurance*, and *resilience*. All these concepts can be boiled down to a core objective—ensuring the ability of Air Force installations to perform mission-essential functions under adverse conditions.

The Air Force asked RAND Project AIR FORCE (PAF) to develop a framework for assessing the resilience of energy delivered to Air Force installations. This report presents the results of that study. In conducting the study, we drew on prior RAND work for the U.S. Department of Defense on capabilities-based planning for energy security (Samaras and Willis, 2013) and for the U.S. Department of Energy on energy resilience as starting points for developing a framework specific to the Air Force's needs. We also spoke with civil engineers on bases, mission support personnel, and mission operators at several Air Force bases.

The concepts described and the framework developed are intended to be scalable, meaning that they can be applied to a single facility, to an Air Force base, or to a mission operation that spans multiple bases. In this report, we focus on the "base" as the default subject, but that is purely for convenience of the discussion. The intended audience for this document includes energy planners at the base level, as well as the higher-level decisionmakers across the Air Force who must make energy assurance investment decisions. In addition to providing a framework for systematically thinking through the question of whether there is an energy assurance problem, this report offers actionable guidance to potential users on how to implement the framework.

The research reported here was commissioned by Edwin Oshiba, Deputy Director of Civil Engineers, Headquarters Air Force A4, and conducted within the Resource Management Program of RAND Project AIR FORCE as part of the project, "Air Force Installation Energy Assurance."

RAND Project AIR FORCE

RAND Project AIR FORCE (PAF), a division of the RAND Corporation, is the U.S. Air Force's federally funded research and development center for studies and analyses. PAF provides the Air Force with independent analyses of policy alternatives affecting the development, employment, combat readiness, and support of current and future air, space, and cyberforces. Research is conducted in four programs: Force Modernization and Employment;

Manpower, Personnel, and Training; Resource Management; and Strategy and Doctrine. The research reported here was prepared under contract FA7014-16-D-1000.

Additional information about PAF is available on our website: www.rand.org/paf/

This report documents work originally shared with the U.S. Air Force on August 31, 2016. The draft report, issued on September 29, 2016, was reviewed by formal peer reviewers and U.S. Air Force subject-matter experts.

Contents

Figures

Tables

Summary

New and evolving U.S. Air Force missions are making mission success more and more dependent on installations in the continental United States (CONUS) and assured access to energy. Especially, mission-essential facilities and installations increasingly require access to electric power to provide mission assurance. In recent years, the Air Force has recognized these issues and has started to increase its focus on what has variously been referred to as *energy security*, *energy assurance*, and *resilience*. All these concepts can be boiled down to a core objective: ensuring the ability of Air Force installations to perform mission-essential functions under adverse conditions.

The Air Force asked RAND Project AIR FORCE to develop a framework for assessing the resilience of energy delivered to Air Force installations. This report presents the results of the study. In conducting the study, we drew on prior RAND work for the U.S. Department of Defense on capabilities-based planning for energy security and for the U.S. Department of Energy on energy resilience as starting points for developing a framework specific to the Air Force's needs. We also spoke with civil engineers on bases, mission support personnel, and mission operators at several Air Force bases.

In completing this work, we focused on three key tasks:

1. **Developing and defining key terms and metrics related to energy assurance.** Many terms are commonly used to describe energy assurance and related concepts, and each of these terms has been given multiple and often inconsistent definitions. We first developed clear and consistent definitions and metrics for key concepts related to energy assurance and used them when working through the framework. While we focus here on electric power, our definitions, concepts, and approach can apply to any energy commodity or other utility.
2. **Describing the electric power architecture status quo at U.S. Air Force installations.** We developed an understanding of electric power system architecture to use as a basis for our analytical framework. This architecture includes both the physical infrastructure and the relevant procedures, authorities, personnel, training, and data as they exist at most CONUS Air Force bases today.
3. **Developing a framework for assessing and improving energy assurance.** Equipped with an understanding of the status quo, as described in Chapter 3, and drawing on established methods of scenario-based planning, as described in Chapter 4, we developed an initial structure for the energy assurance assessment framework, which we then refined with input from base and mission personnel on its usefulness and usability. The energy assurance framework primarily provides a method for base and/or mission personnel to identify gaps in energy assurance that exist at a given base or facility. The framework also provides some guidance for identifying appropriate solutions for different types of problems.

A primary motivation behind this work is that, before investing in potentially expensive installation energy assurance upgrades, the Air Force ought to systematically assess whether there are currently critical gaps. RAND's proposed energy assurance assessment framework is meant to identify such critical gaps. Exercising the framework might lead to the conclusion that the power system architectures currently in place at Air Force installations are, independently or as an enterprise, already capable of handling a wide variety of potential disruptions. If this is the case, no further investments in energy assurance are needed, and resources should be directed elsewhere before expensive upgrades are undertaken. On the other hand, if gaps do exist, the proposed framework can (1) make existing capability gaps and associated risks explicit; (2) help identify the appropriate types of doctrine, organization, training, materiel, leadership and education, personnel, facilities, and policy solutions available to mitigate different types of gaps; and (3) guide informed risk acceptance.

Key Terms and Definitions

To assess and enhance the energy assurance of Air Force installations, we first must be clear about what exactly we mean by "energy assurance." A large number of terms are commonly used to describe energy assurance and related concepts, and each of these terms has been given multiple and often inconsistent definitions. We propose a definition of energy assurance that builds on key elements of these existing definitions:

> *Energy assurance* is the level of access to adequate supplies of energy to support Air Force mission-essential functions.

Also, many terms are commonly applied to concepts that are closely related to energy assurance (indeed, these terms are often used interchangeably). Among these are some that we find useful to discuss as components of energy assurance—*reliability*, *resilience*, and *robustness*—as defined in Table S.1. Assessing energy assurance also requires a clear understanding of *requirements* and *capabilities*; both terms are defined in Table S.1.

Metrics

Metrics provide a common language that energy users and providers can use to communicate requirements and capabilities and agree on appropriate actions. Metrics are needed to understand whether problems might arise in the face of certain disruption scenarios and the extent and criticality of the problems. In this report, we propose a number of different types of metrics. *Requirement metrics* and *capability metrics* are used by users and providers, respectively, to communicate with each other. These metrics may or may not be the same but will typically be related such that they can be combined to form *performance metrics*, which describe the magnitude of any gap that exists between the two. The last type of metric we propose is the *tracking metric*. Tracking metrics do not necessarily describe mission-assurance–related

Table S.1. Definitions of Energy Assurance Components

Component of Energy Assurance	Definition	Additional Explanation
Reliability	The confidence in the actual power characteristics provided to a point in the system	This definition covers both the quality and the amount of power provided.
Resilience	The ability of a system to withstand and recover from a disruption	Resilience is often discussed in the context of high-consequence, low-probability events, such as natural disasters or determined attacks.
Robustness	The ability to adequately meet power requirements across multiple scenarios	A system may be subjected to any number of different types of disruption scenarios. No system can be robust across all imaginable scenarios.
Requirements	The power characteristics a user needs from a supplier at a given point in the system	Failing to meet the requirement should have some mission impact; otherwise, the requirement is arbitrary.
Capabilities	The ability of the supplier to provide power characteristics at a given point in the system	Capabilities describe what can be accomplished, unrelated to what is used or required.

requirements or capabilities but are useful for distinguishing among alternative architectures and solution options once problems have been identified. See Table S.2 for our proposed list of energy assurance metrics.

Reliability can be assessed by looking at the quality and level of power supplied (i.e., critical load not served, nominal load not served, total harmonic distortion, and voltage sags and swells). Resilience can be assessed by looking at the level of reliability when facing a disruption and, in the case of degraded performance, the time to restore mission functions and nominal operations. Robustness can be assessed by looking at system performance in reliability and resilience across a wide range of possible scenarios. It is important to note that the same metric might be viewed as a reliability metric as opposed to a resilience metric, depending on the perspective of the user. Reliability is defined upstream of any given point in the system, and resilience is defined downstream.

In addition to proposing a set of useful metrics, we surveyed the literature for attributes of "good" metrics to guide our selection process. These attributes are meant to provide a way for users of the RAND framework to systematically select new metrics:

- validity
- policy relevance
- maturity
- operational usefulness
- resource intensiveness.

Table S.2. Energy Assurance Metrics

Metric	Requirement	Capability	Performance	Tracking	Typical Units
Amount of power					
Power supplied		X			kW
Critical demand	X				kW
Critical load not served			X		kW
Nominal demand	X				kW
Nominal load not served			X		kW
Power quality					
Total harmonic distortion	X	X			Deviation (%)
Voltage sags, swells	X	X			Deviation (%)
Gap between required and actual power quality			X		Deviation (%)
Restoration					
Time to restore critical functions	X	X			Seconds to days
Gap between required and actual restoration time for critical functions			X		Seconds to days
Time to restore nominal operations	X	X			Seconds to days
Gap between required and actual restoration time for nominal operations			X		Seconds to days
Outage cost				X	$

RAND Energy Assurance Framework

We began by gaining an understanding of the status quo electric power system architectures, as described in detail in Chapter 3. Drawing on established methods of scenario-based planning to define the critical dimensions of event-based scenarios, we developed an initial structure for the energy assurance assessment framework. We presented our initial hypotheses about the framework to and discussed them with base and mission personnel, honing in on the elements they considered useful and actionable—and those that were less so. We then refined this initial framework to sharpen definitions, sequence analyses, and determine relationships among the components. These changes were guided by our objective of creating an easy-to-use, logical, and replicable approach to assessing energy assurance with sufficient flexibility to apply across the entire enterprise.

The RAND framework provides a way for civil engineers on bases and mission stakeholders to assess whether gaps exist and of what kinds between energy capabilities and energy requirements at a given base or installation under a range of scenarios. If such gaps exist, the framework then provides a structured way to identify viable response options that work well across scenarios to close gaps or otherwise mitigate risks. The framework is focused on an individual base or installation but could be used at higher levels to support integrated analysis and decisionmaking among base commanders, civil engineers on base, and mission owners[1] in their respective roles of making installation-level investment and operational decisions to support missions.

The RAND framework guides mission owners and civil engineers on base through a series of steps. As shown in Figure S.1, mission owners and civil engineers would gather information to characterize the essential features of a base's electrical system architecture, capabilities, and requirements. They would then use this information to assess potential outcomes associated with gaps between capabilities and requirements (performance) when the base or its environs are subjected to an event-driven scenario. This process would be repeated for each scenario. In Figure S.1, the green boxes represent information that is independent of scenarios. The orange box represents scenarios disruptive to "normal" operations. The gray boxes represent information conditional on scenarios. The yellow diamonds represent modeling and simulation efforts, or some form of discussion-based exercises needed to assess capabilities and outcomes associated with degraded capabilities.

In the next steps of the framework shown in Figure S.2, mission owners and civil engineers would identify available response options across scenario outcomes; analyze these potential solution options; implement solutions that perform well across scenarios; and document, accept, and periodically review risk mitigation and acceptance decisions.

Scenarios

When the base system architecture is subjected to the stress of conditions associated with an externally driven scenario, base capabilities could be compromised, disrupted, degraded, or fully disabled. Scenarios can serve as a starting point for stress-testing bases to disruptions in power that could affect mission-essential capabilities over the short and longer terms. Scenarios are outside the control of the base or the Air Force, although the base leaders or others in the Air Force could mitigate the consequences of each scenario for the base's mission-essential energy.

---

[1] We broadly define mission owners as commanders of organizations responsible for some kind of operational outcome. A mission could be fighter pilot training; operating and maintaining unmanned aerial systems; or operating a radar installation, command headquarters, or tenant unit for another service or government agency.

Figure S.1. Energy Assurance Framework: Assess Performance for One Scenario

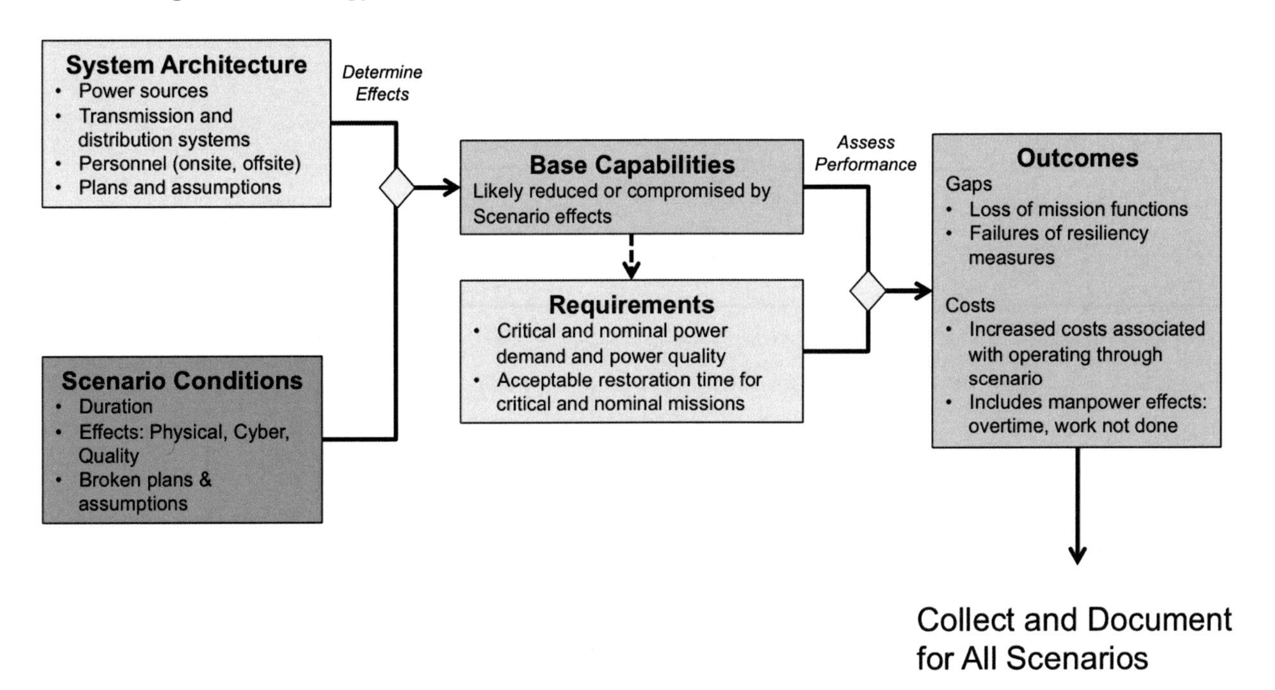

Figure S.2. Energy Assurance Framework: Assess Performance Across Scenarios

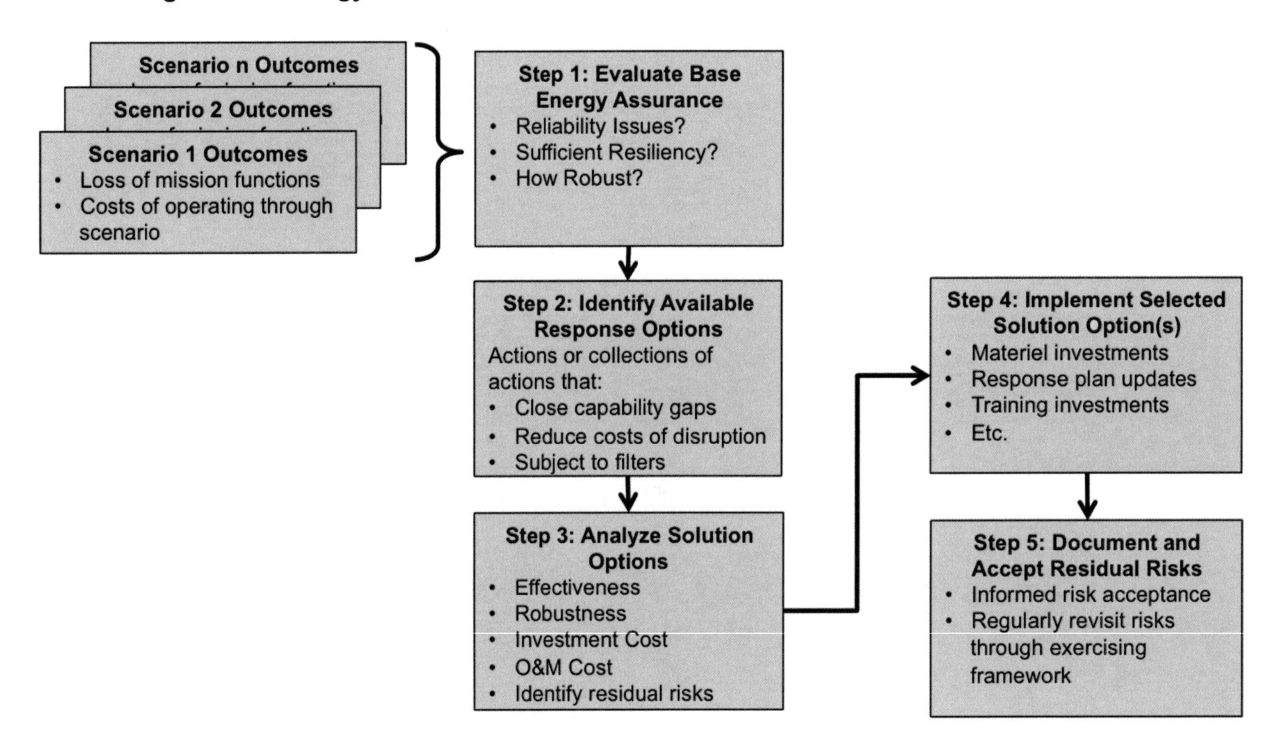

We define event-driven scenarios as sets of conditions triggered by external physical occurrences, such as storms, or by the disruptive and nefarious efforts of determined adversaries. In the context of energy assurance, we assume that the Air Force has operational plans and contingencies in place for common scenarios, such as routine equipment failures or failure due to normal wear and tear. Here, we instead focus on scenarios intended to stress operational plans and assumptions to their breaking point as a way of probing the range of vulnerabilities that installations may have but not know they have. Each event-driven scenario involves a complete loss of external power to the base, which may last for minutes or months. Table S.3 describes the five conditions used to define an event-based scenario. Each condition is varied by extent, such as a length of time the condition holds or the magnitude of geographic area that experiences such a condition.

Table S.4 describes five scenarios we recommend bases use to assess the resilience of their energy systems.

In developing the scenarios presented in Table S.4, we sought a relatively small set of scenarios that covers as much of the (plausible) uncertainty space as possible. Figure S.3 presents this coverage visually, with the five selected event-driven scenarios labeled E1 through E5.

Before investing in solutions to problems identified by evaluating performance against event-driven scenarios using the framework, decisionmakers should also consider possible changes or shifts in future conditions that could constrain proposed solutions or render them obsolete. Table S.5 presents a sampling of changes in future conditions that decisionmakers should consider.

Table S.3. Structure of Event-Driven Scenarios

Condition	Description	Extent
Duration	Period over which event causes outages or other disruptions	Hours, days, weeks, months
Physical effects	Weather or terrorist events physically damage equipment or disrupt operations	Base, local, regional
Cybereffects	Internet or information technology systems are compromised or inaccessible	Base, local, regional
Power quality effects	Voltage sags or swells that damage or otherwise degrade sensitive equipment	Present, not present
Broken assumptions and plans	Critical system architecture elements break down in unexpected ways	Backup systems fail to turn on, loss of access to base, etc.

Table S.4. Proposed Initial Set of Event-Driven Scenarios

	Scenario E1 Delta	Scenario E2 Joplin	Scenario E3 Icestorm/Sandy	Scenario E4 Cyberattack	Scenario E5 Sandy + Cyber
Duration	12 hours	3–7 days	2 weeks	1 month	3 months
Physical effects	Base	Local	Regional	None	Regional
Cybereffects	None	None	None	Base	Regional
Power quality	Present	Not present	Not present	Present	Present
Scenario narrative	A lightning strike on base power line causes local fire and power quality event	High winds create large debris field on base and in surrounding community	An ice storm severely damages power lines and trips relays or a hurricane causes severe flooding and wind damage; off-base communications, landlines down	An adversary attacks information technology and backup power systems on the base and also physically targets critical nodes in the power system, cutting the power grid	Combination of Scenarios E3 and E4, where an adversary launches a targeted cyberattack following or in the midst of a Sandy-like disaster
(Sample) Broken plans and assumptions	Instruments and other equipment cannot restart following event; data unavailable	Off-base support personnel and fuel service unavailable because of downed lines and debris; communications capabilities lost	Off-base support personnel and fuel service unavailable because of downed lines and debris; communications capabilities lost	Instruments and other equipment cannot restart following event; data unavailable; loss of all communications	Off-base support personnel and fuel service unavailable; loss of access to data; loss of communications

Figure S.3. Diverse Coverage of Scenario Space

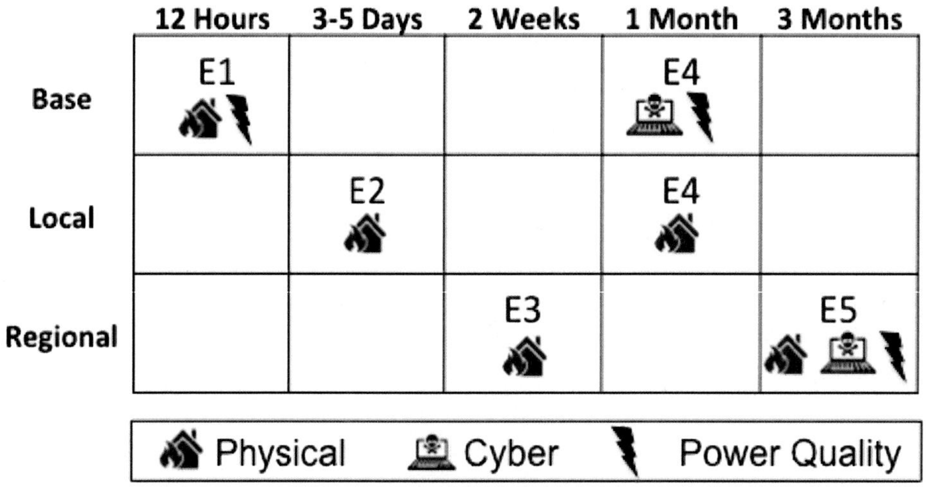

	12 Hours	3-5 Days	2 Weeks	1 Month	3 Months
Base	E1			E4	
Local		E2		E4	
Regional			E3		E5

Physical Cyber Power Quality

Table S.5. Sample Changes in Future Conditions

Solution Category	Changes in Future Conditions to Consider	Potential Drivers
Operational and personnel changes that require manpower (with specialized skills or otherwise)	• Drastically reduced military budgets • Large-scale deployment of Air Force personnel overseas • Difficulties retaining or growing the civil engineer career field	• Global shifts in defense priorities • The United States is engaged in a major war • Demand for civil engineers and energy technologists goes up, making it harder for the Air Force to recruit and retain the best and the brightest
Investments in equipment, technology, infrastructure, facility recapitalization	• Changed prime and backup power supply mix: no nuclear; renewables only (no diesel backup) • Changed grid configuration: regional or national grid with increased distributed generation sources, or increased penetration of microgrids; no national/regional grid—only distributed generation and microgrids	• U.S. nuclear plants shut down after major accident; no exemptions granted for the Air Force • Carbon legislation passed; increased penetration of renewables and other distributed generation • Public consensus to fully phase out all uses of fossil fuels and nuclear power; renewables only in power mix; maturation of power storage technologies

Evaluating Base Energy Assurance

The framework guides a user through the process of assessing performance (i.e., gaps between requirements and capabilities) against each scenario of interest first. Outcomes of an individual scenario come in two forms: (1) the loss of mission functions or failure of resiliency measures resulting from gaps between requirements and capabilities and (2) total marginal outage costs associated with operating through a scenario, including manpower effects, such as overtime or opportunity costs associated with other work not done. After working through the framework cycle for each scenario, the next step is to synthesize base performance across the individual scenarios. In this step, users should first identify commonly occurring gaps across scenarios and dig deeper to identify the underlying causes. After identifying classes of problems affecting a particular base by running through different scenarios, the next step is to identify appropriate solutions (considering their relative effectiveness, robustness, costs, and the residual risks associated with implementing them). We provide an approach (in Appendix C) for identifying and sifting through viable solutions using a filtering process that accounts for technical constraints and requires the user to clearly understand and document the risks associated with each solution. Whichever solution option is ultimately chosen, the questions of which risks have been addressed and which risks are likely to remain unaddressed will need to be documented from multiple perspectives: mission owners, civil engineers, and headquarters. Further, a mission risk that exists at one base potentially could be mitigated by investing in another base that supports the same mission. The ultimate goal of the framework is to help decisionmakers take an enterprise view of energy assurance when making risk-informed decisions about whether to invest in energy assurance upgrades and, if so, in what ways.

Key Findings Based on Discussions with Air Force Personnel

Air Force installations in CONUS generally have similar electrical system architectures. In all cases, the local utility provides power with some reliability. Commercial power, often with only one or two lines and one or two substations, feeds the entire base. On-base resilience in the event of a power grid disruption is primarily provided by generators (owned, operated, and maintained by civil engineers on base) and uninterruptible power supplies (typically owned and operated by mission owners with maintenance support provided by off-base contractors[2]). Continuity of operations plans ensure that the mission continues even if a base is unable to operate fully (e.g., by moving the mission to another location). Access to fuel for generators is a key component of resilience capabilities. Scenarios that combine outages with constrained fuel delivery and/or reduced access to off-base personnel and parts can severely disrupt missions. While civil engineering organizations might be aware of single points of failure (e.g., critical communications nodes or aging transmission lines that serve large portions of a base), resilience capacities are often determined within organizational silos and not developed bearing the resources of the whole base in mind. This limits opportunities to develop novel solutions and visibility into dependence on single points of failure. For instance, civil engineers are generally responsible for planning, programming, and overseeing base activities for military construction projects at their installations (Air Force Instruction 32-1023, 2015). However, when a new facility is planned for construction at a base, it is unclear whether due consideration is given to how that facility might affect prioritization of backup resources on the base.

Mission owners often define or communicate requirements (e.g., acceptable mission downtime) poorly. Even when requirements are specified, it is not always clear what drives them or how they tie into mission goals. At times, requirements seem to be based on what can be achieved rather than being rooted in true mission goals. Mission-owner power requirements are often not communicated clearly to civil engineers in a timely fashion, before a disruptive event has taken place.

Base personnel are typically well aware of the consequences of exposure to previous events for base capabilities and are ready to take necessary actions to cope. On the other hand, there is limited understanding of whether and how base capabilities might degrade in response to disruption scenarios that the base has not yet experienced.

---

[2] On rare occasions, base civil engineers might also own and operate uninterruptible power supplies.

Recommendations

- Mission owners should define energy requirements up front and clearly communicate them to the civil engineer. Currently, the prioritized asset list drives civil engineering activities during outages, but civil engineers are often forced to react to demands in real time.
- Energy requirements should be clearly tied to mission goals and needs. If an unmet "requirement" seemingly has no implication for the mission, it is not a real requirement.
- Assessment of electric power requirements should account for interdependencies between electricity and other mission-critical infrastructures, such as water.
- To the extent possible, installation energy planners and mission owners who rely on assured access to electric power should use metrics, such as the ones proposed in Chapter 2, to articulate requirements, capabilities, gaps between the two, and the implications of any gaps.
- Operators and planners should invest in better understanding the effects of exposure to scenarios that have not yet not occurred. A thorough assessment would require physical testing and modeling and simulation efforts. But simple tabletop exercises, such as the one outlined in Appendix B, which focuses on asking questions that reveal implicit biases about how systems and people operate on a base, can yield critical insights into the extent to which an installation is truly prepared to face disruptions of different types. Such exercises can also raise awareness of important trade-offs between increased efficiency and reduced security. For instance, some investments, such as remote monitoring of generators, increase resilience to flood scenarios but may diminish capabilities in cyberattack scenarios. Much like having a variety of generator makes and models, not putting all backup systems into a single basket can increase resilience to a variety of scenarios.
- Decisionmakers should look across missions at a given base and across bases that support a particular mission before investing in energy assurance upgrades. Taking a holistic look can help ensure that requirements are not identified in isolation, capabilities are not developed in silos, and resources are efficiently used. For instance, a mission risk present at one base could be mitigated by investing in another base that supports the same mission and is better suited for cost-effective resilience upgrades.
- In picking solutions to implement at different bases, decisionmakers should consider possible changes or shifts in future shifts driven by changes in policy, environmental, or economic conditions that could constrain or render obsolete certain solutions in addition to analyzing costs, risks, and other base-specific considerations. Using a filtering process, such as the one outlined in Appendix C, coupled with a careful consideration of longer-term, slower-moving changes, such as those described in Chapter 4, can help increase the likelihood of solutions chosen today remaining applicable and effective in an uncertain future.

Acknowledgments

We are grateful for the extraordinary support of our sponsor, Edwin Oshiba, Deputy Director of Civil Engineers, Headquarters Air Force A4C. We would also like to thank our points of contact at the Office of the Assistant Secretary of the Air Force for Installations, Environment and Energy, Doug Tucker and Mike Wu, who provided guidance and support and helped us make valuable connections with Air Force subject-matter experts throughout the course of the project.

We would like to thank various people across the Air Force for taking the time to engage with us and providing information that served as the basis for our analysis. In no particular order, these individuals include Tarone Watley, Daniel Soto, and Deven Volk at the Air Force Civil Engineering Center; Lt Col Kevin Osborne (Base Civil Engineer), Maj Sean Stapler, Kevin Rasmussen, and the civil engineering team and representatives of the 480th Intelligence, Surveillance, and Reconnaissance Wing, the 27th Intelligence Squadron, and the 497th Intelligence, Surveillance, and Reconnaissance Group at Langley Air Force Base (AFB); Lt Col Bryan Opperman (Base Civil Engineer) and the civil engineering team at Minot AFB; Lt Col Greg Mayer (Base Civil Engineer) and the civil engineering team at Luke AFB; the civil engineering team and representatives from the 9th Reconnaissance Wing, the 548th Intelligence, Surveillance, and Reconnaissance Group, and the 7th Space Warning Squadron at Beale AFB; the civil engineering team at Joint Base McGuire-Dix-Lakehurst; and Michelle Linn at Air Force Space Command A4C.

Within RAND, the authors are thankful to Obaid Younossi for his program leadership and project guidance and to Nidhi Kalra and Henry Willis for their critical review of the work as part of a red-team exercise. We would like to also thank Endy Daehner and Aimee Curtright for their peer reviews of this document.

Abbreviations

ABP	assumption-based planning
ACC	Air Combat Command
AFB	Air Force base
AFCEC	Air Force Civil Engineering Center
AFI	Air Force instruction
AFIMSC	Air Force Installation and Mission Support Center
AFSOUTH	Air Forces Southern
BCE	base civil engineer
CHP	combined heat and power
CONUS	continental United States
COOP	continuity of operations plans
DCIP	Defense Critical Infrastructure Program
DoD	Department of Defense
EMCS	energy monitoring and control system
EV	electric vehicle
FB	facilities board
FY	fiscal year
HAF	Headquarters Air Force
IC	installation commander
IPCC	Intergovernmental Panel on Climate Change
IT	information technology
LNG	liquefied natural gas
MAJCOM	major command
MDG	medical group
MILCON	military construction
MO	mission owner
MSG	mission support group
O&M	operations and maintenance
OSD	Office of the Secretary of Defense
PDU	power distribution unit
REDI	Resilient Energy Demonstration Initiative
RMS	root-mean-squared
RPIE	Real Property Installed Equipment
SAF	Secretary of the Air Force
SAIDI	System Average Interruption Duration Index

THD	total harmonic distortion
Tmf	time to restore mission functions
Tn	time to resore nominal operations
UPS	uninterruptible power supply
USAF	U.S. Air Force
WG/CC	Wing Commander

1. Introduction

The success of all critical U.S. Air Force missions depends on access to some form of energy. Traditionally, energy for the Air Force has meant jet fuel, which accounts for about 80 percent of total Air Force energy consumption. However, the Air Force's new and evolving missions are becoming more and more dependent on installations in the continental United States (CONUS). Remotely piloted aircraft; intelligence processing, exploitation, and dissemination; networked real-time communications for command and control; and space and cyber missions are all growing rapidly, and the success of all depends on uninterrupted land-based operations. In this evolving landscape, mission-essential facilities and installations require, among other things, continuous and assured access to electricity. In recent years, the Air Force has recognized these issues and has started to increase focus on what has variously been referred to as *energy security*, *energy assurance*, and *resilience*—fundamentally, ensuring and protecting the ability of installations to perform mission-essential functions under adverse conditions. With the establishment of the Air Force Office of Energy Assurance, along with a number of initiatives focused on evaluating demonstration projects designed to increase base energy resilience, the Air Force is increasing focus on this area.

Most CONUS bases are entirely dependent on the commercial U.S. grid to provide primary electric power, but all facilities that are considered mission-essential typically have some form of backup power system, almost always diesel-fueled generators. Further, most computer systems and other sensitive and critical systems are connected to uninterruptible power supplies, both to ensure consistent power quality and to cover operations during power outages until the generators begin working. Air Force mission functions that are based in these facilities have continuity of operations plans (COOPs) that go into effect when significant disruptions occur. These may include transferring mission functions to a backup location, switching to alternative communications modes, or engaging in other activities that mitigate the disruption. All these measures are specified in law or policy. These measures are generally targeted toward the relatively high-frequency, low-impact kinds of disruptions that we are used to preparing for: grid outages that last for minutes to hours. However, the Air Force does not yet have a systematic or preemptive approach for evaluating energy assurance at installations and determining how disruptions might affect missions, especially under more-severe disruptions than current plans take into account.

Research Objective and Approach

The Air Force asked RAND Project AIR FORCE to develop a framework for assessing the resilience of energy delivered to Air Force installations. This report presents the results of the study.

In conducting the study, we drew on prior RAND work for the U.S. Department of Defense (DoD) on capabilities-based planning for energy security and for the U.S. Department of Energy on energy resilience as starting points for developing a framework specific to the Air Force's needs. We also spoke with civil engineers, mission support personnel, and mission operators at several Air Force bases (AFBs). Further, we visited Beale AFB and Langley AFB to better understand the status of energy assurance and to help develop and test our framework.[3]

The concepts described and the framework we developed are intended to be scalable, meaning that they can be applied to a single facility, to an AFB, or to a mission operation that spans multiple bases. In this report, we focus on the base as a starting point. The concepts introduced in this report can certainly be applied at the enterprise level, but doing so would require digging deeper into a few additional questions, for example: How do different bases work together to support a particular mission? How might the presented metrics need to be modified or adapted when used in an enterprisewide context? Who would conduct enterprise-level analyses and oversee investment decisions? These questions are outside the scope of this work, but the framework should set the stage for future work that is aimed at answering them.

The work focused on several topic areas, and the report is structured along similar lines:

1. **Develop and define key terms and metrics related to energy assurance.** As discussed earlier, many terms are commonly used to describe energy assurance and related concepts, and each of these terms has been given multiple and often inconsistent definitions. We first developed clear and consistent definitions for key concepts related to energy assurance and used them when working through the framework. We also surveyed the literature on energy assurance metrics (see Appendix A) and, in combination with our discussions with base personnel, identified and clearly defined a small set of metrics that we believe will be most generally useful to planners and operators. These terms and metrics, their definitions, and the context in which they are applied are the subjects of Chapter 2.
2. **Describe the electric power architecture status quo at Air Force installations.** Chapter 3 covers, in general terms, the existing electric power system architecture as it exists at most Air Force installations in CONUS. This architecture includes not only physical infrastructure but also relevant procedures, authorities, personnel, training, and data. Understanding the existing architecture provides a baseline and helps ground the discussion of the framework.
3. **Develop a framework for assessing and improving energy assurance.** The energy assurance framework primarily provides a method for base and/or mission personnel to

---

[3] The visit to Beale AFB occurred as part of a site visit conducted by the Resilient Energy Demonstration Initiative (REDI).

identify gaps in energy assurance at a given base or facility. The framework also provides some guidance for identifying appropriate solutions for different types of problems. We applied the principles of capabilities-based planning and decisionmaking under uncertainty to develop a small, manageable set of plausible disruption scenarios that would stress Air Force electric power systems across a wide range of conditions. These scenarios, along with the existing electric power architecture of the base, served as inputs to the framework, interacting to determine (likely degraded) base capabilities. We compared these against the requirements provided by mission functions, an analysis that uses the set of metrics described in Chapter 2 to identify energy assurance gaps. The framework and its steps are described in Chapters 4 and 5, with Appendix B providing more detailed guidance on implementation. Identifying specific energy assurance solutions, particularly at a base or mission level, is a complex task requiring detailed analysis well beyond the scope of this report. However, Appendix C provides guidance on how to identify appropriate solutions for different types of gaps.

4. **Provide findings and recommendations.** Although these were not the focus of this project, Chapter 6 discusses some key findings and recommendations that grew out of the site visits and discussions with Air Force personnel.

2. Energy Assurance Definitions and Metrics

To assess and enhance the energy assurance of Air Force installations, we must first be clear about what exactly we mean by *energy assurance*. As noted earlier, many terms are commonly used to describe energy assurance and related concepts, and each of these terms has been given multiple and often inconsistent definitions. In the first part of this chapter, we will define the terminology that will be used throughout this report. Note that, while our work has focused on electric power, the definitions, concepts, and approach laid out in this report can apply to any energy commodity or other utility.

The second part of the chapter will focus on installation energy assurance metrics, the careful selection of which is necessary for assessing the status of energy assurance and the nature and magnitude of gaps between requirements and capabilities. Many metrics have been proposed for measuring the reliability, resilience, and other properties of electric power systems; here, we will narrow these down to a handful of simple parameters that should be broadly useful and measurable across Air Force installations and missions. Because particular bases or missions may have specific needs for other or more complex measures, we also provide some guidance for selecting which among the many metrics available are likely to be the most useful.

Energy Assurance

The DoD defines *mission assurance* as

> A process to protect or ensure the continued function and resilience of capabilities and assets—including personnel, equipment, facilities, networks, information and information systems, infrastructure, and supply chains—critical to the performance of DoD [mission-essential functions] in any operating environment or condition. (DoD, 2012a)

The purpose of this work is to develop a process for examining (and ultimately improving) the ability of Air Force installations and ground-based mission functions to support mission assurance by ensuring the continued function and resilience of capabilities and assets that depend on energy (in the form of electric power), in any operating environment or condition. We believe the term *energy assurance* best describes this overarching concept and, therefore, adopt it here.

The 2010 Quadrennial Defense Review made use of the term *energy security* for a very similar concept, defining it as "having assured access to reliable supplies of energy and the ability to protect and deliver sufficient energy to meet operational needs" (DoD, 2010, p. 87). A slightly modified version of this definition appeared in the Fiscal Year (FY) 2012 National Defense Authorization Act, which stated that energy security is "having assured access to reliable supplies of energy and the ability to protect and deliver sufficient energy to meet mission essential requirements" (Public Law 112-81, 2011). Examining the components of this second definition reveals that it identifies a goal—"sufficient energy to meet mission essential requirements"—and a means toward achieving that goal—"having assured access to reliable supplies of energy and the ability to protect and deliver."

We propose a definition of energy assurance that builds on ideas and language found in these existing definitions:

> *Energy assurance* is the level of access to adequate supplies of energy to support Air Force mission-essential functions.

This definition has a few key elements:

- "Level of access" implies that energy assurance is something that can be measured, rather than a binary concept that either holds or does not hold. Needing to assess the "level" of access, in turn, requires the existence or development of useful metrics and measures. The level of access itself, properly measured, can be understood as a capability of the system.
- "Adequate supplies" implies that there is a requirement, an agreement on what level of access qualifies as adequate for a given mission or base function. When the requirement (or set of requirements) is met by existing capabilities, adequate supplies have been achieved. When the requirement exceeds the capability, a gap exists, and energy assurance is not fully achieved.
- "Mission-essential" implies that there is a consequence for failing to meet requirements. If a capability gap is observed but has no mission impact, it is very likely that the requirement is arbitrary or excessive and should be revisited.

Each of these elements—metrics, requirements, and mission-essentiality—is a central component of the framework for assessing energy assurance, as Chapter 5 will describe in detail.

Reliability, Resilience, and Robustness

Many terms are commonly applied to concepts that are closely related to energy assurance (and, indeed, are often used interchangeably). Among these are some that we find useful to discuss as components of energy assurance: these are *reliability*, *resilience*, and *robustness*. We give definitions here for how these terms will be used throughout the report.

Reliability is a key component of energy assurance. The North American Energy Reliability Corporation views reliability as the "degree to which the performance of the elements of that system results in power being delivered to consumers within accepted standards and in the amount desired" (North American Electric Reliability Corporation, 2012; see also Osborn and Kawann, 2001; North American Electric Reliability Corporation, 2013; and Hirst and Kirby, 2000). Drawing from this, we chose the following definition:

> *Reliability* is the confidence in the actual power characteristics provided to a point in the system.

This broad definition covers both the quality and the amount of power provided. Reliability can be increased in a number of ways, including component design (e.g., more-reliable individual components), system design (e.g., redundant components, multiple pathways), and provider response and recovery capability (e.g., automation, sensors, and repair and maintenance capabilities). Ultimately these actions simply translate into power characteristics provided at a point.

Resilience is defined for a particular system in terms of particular disruption scenarios. Drawing from the vast literature on the topic,[4] we chose the following definition:

> *Resilience* is the ability of a system to withstand and recover from a disruption.

Resilience is often discussed in the context of high-consequence, low-probability events, such as natural disasters or determined attacks (Watson et al., 2014). We adopt a similar perspective in this work, focusing on events that fall outside the common short-duration power outages that all installations account for in their infrastructure designs. Traditional risk assessment approaches are not directly useful when it comes to planning for such events because of the lack of confidence in the associated probabilities. Capability-based, or scenario-based, planning is the alternative approach we use to explore the space of potential disruptions and systematically assess requirements and capabilities under different conditions. Resilience generally depends on actions taken by power receivers (whether an end user or not). These actions could include securing alternative sources of electric power (e.g., uninterruptible power supplies [UPSs], generators, batteries), implementing alternative procedures (e.g., pen and paper instead of computers), reducing consumption, etc. The need for resilience arises when nominal system conditions do not hold for some finite time, through an equipment failure or other disruption.

Under these definitions, reliability and resilience are concepts that can vary with perspective. For example, one stakeholder, say the base civil engineer, is responsible for ensuring that power requirements are met for various mission functions housed at the base. The commercial power company provides electric power to the base with a certain reliability, which is largely out of the control of the base civil engineer. When the power from that source is lost during an outage, the base civil engineer is responsible for providing resilience, which may involve turning on backup power generators, reducing unnecessary power consumption, and/or taking other measures to ensure that mission-essential functions on the base still receive whatever power they need to continue operating. In general, these resilience measures represent investments made on the base. To understand the proper level of investment to make in resilience capabilities, the civil engineers and other base stakeholders need to understand, among other things, the reliability of

---

[4] See, for example, Air Force Space Command, 2013; Allenby and Fink, 2005; Anderies et al., 2013; Committee on Increasing National Resilience to Hazards and Disasters, Committee on Science, Engineering, and Public Policy, and The National Academies, 2012; Gunderson, 2000; Haimes, 2009; Holland, 2013; Holling, 1973; Ibanez et al., 2016; Jennings, Vugrin, and Belasich, 2013; Masten, 2009; Presidential Policy Directive 21, 2013; Rutter, 2008; Sandia National Laboratories, 2013; Sutcliffe and Vogus, 2003; Trivedi, Kim, and Ghosh, 2009; University of Kansas Information and Telecommunication Technology Center, 2014; Walker et al., 2004; Watson et al., 2014; Wei and Ji, 2010; and Wieland and Wallenburg, 2013.

the power received from the commercial provider and the energy requirements of the mission-essential functions that are provided on their base.

A mission owner, sitting in a building on the base, has a different perspective. In this case, it is the power to the building that is provided with a certain reliability by the combination of the commercial power company and the civil engineers on the base. If this power is lost, perhaps because a backup generator fails, the mission owner needs to be resilient to that, perhaps by relying on UPSs for some time, by transferring the mission function to another base or, for example, reverting to using pencil and paper.

A system may be subjected to many different types of disruption scenarios. Drawing from the literature on robust decisionmaking (see, for instance, Lempert, Popper, and Bankes, 2003, and Lempert et al., 2013),

> *Robustness* is the ability of a base, mission, or other unit of command to adequately meet power requirements across a wide range of possible scenarios.

The robustness of the energy system is assessed through evaluation of performance metrics across multiple disruption scenarios. No system can be robust across all imaginable scenarios, and attempting to make one so would certainly involve exorbitant costs to account for extremely rare events. But in general, the wider the range of scenarios against which a base (or any system of study) performs well across metrics of interest, the better it can cope with surprise and, consequently, the more robust it is. Robustness can be improved by building capabilities that are not vulnerable to the same failure modes or conditions (e.g., two types of backup power). For a system that performs poorly under certain scenarios, decisions must be made about whether to mitigate the associated risk through increased investment or changes in operations or whether to accept that risk.

None of the conditions of reliability, resilience, or robustness can be assessed properly without understanding and being able to articulate energy *requirements* and *capabilities* clearly.

Energy Requirements and Capabilities

For our purposes, *requirements* consist of the power characteristics a user needs from a supplier at a given point in the system. *Capabilities* describe the ability of the supplier to provide power characteristics at a given point in the system. At any given time, the actual power provided may be less than the full capability: The capability describes what can be accomplished, unrelated to what is used or required.

Under normal conditions, the capability of a system is determined by its architecture. In the case of an Air Force base, this architecture would encompass every resource available to support the provision of power to users, including the commercial power lines entering the base, substations, transmission lines and distribution systems on the base, backup power generation equipment, the trained personnel (local or remote, Air Force civilian or contractor) who maintain and operate these systems, maintenance contracts and policies, and fuel storage and delivery agreements. Understanding the full capabilities of a system requires knowledge of all the components of the system architecture and how they fit together.

External or unusual conditions can affect (reduce, eliminate) capabilities and stress the system. If a system is not very resilient or is placed under sufficient stress, capabilities can be reduced to the point that not all of the requirements can be met. It is these "capability gaps" that the energy assurance framework is intended to identify.

Capabilities, requirements, and gaps cannot be evaluated in a meaningful way without the use of metrics. The next section describes the set of basic metrics we propose using with the framework and an approach for identifying appropriate additional metrics when needed.

Energy Assurance Metrics

Metrics provide a common language that energy users and providers can use to communicate requirements and capabilities and agree on appropriate actions. Metrics are needed to understand whether problems might arise in the face of certain disruption scenarios and to understand the extent and criticality of the problems.

In this report, we use a number of different types of metrics. Users and providers use *requirement metrics* and *capability metrics*, respectively, to communicate with each other. These metrics may or may not be the same but will typically be related such that they can be combined to form *performance metrics*, which describe the magnitude of any gap that exists between the two. As an example, a mission function may have a minimum demand for power to operate critical systems of 100 kW. Meanwhile, the base has a nominal capability to supply power, also described in kilowatts, that exceeds this demand. During a power outage, the capability to supply power could be temporarily reduced to 0 kW. At this point, a gap exists, and the size of that gap is described by a performance metric called *critical load not served*: the power required less the capability to provide power (100 kW in this example).

The last type of metric we propose is the *tracking metric*. Tracking metrics do not necessarily describe mission-assurance–related requirements or capabilities but are useful for distinguishing among alternative architectures and solution options once problems have been identified. Cost is an obvious example of a tracking metric and the only one in this category we propose in this work. Multiple architectures may be able to achieve the same level of mission assurance, but if one is much more cost-effective than another, this is well worth knowing.

We developed a candidate set of potential energy assurance metrics by (1) surveying existing literature for metrics commonly used to evaluate energy system performance and (2) identifying metrics that would be relevant specifically to Air Force installations.[5] From this list, we selected

---

[5] See, for example, Bollen, 2003; Bompard, Napoli, and Xue, 2010; Greene and Lancaster, 2006; International Electrotechnical Commission, 2008; Keogh and Cody, 2013; Kueck et al., 2004; Martínez-Anido et al., 2012; McCarthy, Ogden, and Sperling, 2007; Pillay and Manyage, 2001; Roe and Schulman, 2012; Rouse and Kelly,

the small set of metrics shown in Table 2.1. These metrics cover three categories: amount of power required and supplied, quality of power required and supplied, and duration of a disruption. We also included outage cost as a tracking metric.

We define these metrics as follows:

- ***Critical demand*** is a requirement metric describing the minimum level of power needed at a point (typically a building) to ensure the ability to perform mission-essential functions.

Table 2.1. Energy Assurance Metrics

Metric	Requirement	Capability	Performance	Tracking	Typical Units
Amount of power					
Power supplied		X			kW
Critical demand	X				kW
Critical load not served			X		kW
Nominal demand	X				kW
Nominal load not served			X		kW
Power quality					
Total harmonic distortion	X	X			Deviation (%)
Voltage sags, swells	X	X			Deviation (%)
Gap between required and actual power quality			X		Deviation (%)
Restoration					
Time to restore critical functions	X	X			Seconds to days
Gap between required and actual restoration time for critical functions			X		Seconds to days
Time to restore nominal operations	X	X			Seconds to days
Gap between required and actual restoration time for nominal operations			X		Seconds to days
Outage cost				X	$

2011; Sandia National Laboratories, 2014; Teodorescu and Liserre, 2011; Vijayaraghavan, Brown, and Barnes, 2004; Voorspools and D'Haeseleer, 2004; Yeddanapudi, 2012.

- ***Nominal demand*** is a requirement metric describing the minimum level of power needed at a point to conduct normal operations. Nominal demand will always be greater than or equal to critical demand and will include power for computers, lighting, and other equipment in noncritical areas. These systems perform functions that are not critical during a short disruption but the loss of which will create stress during a long disruption.
- ***Power supplied*** is a capability metric describing the amount of power that the electrical system can provide to a point at any given time. This metric can be used to measure power supplied both during normal operation and during a power-disruption scenario.
- ***Critical load not served*** is a performance metric derived by subtracting the power supplied from the critical demand. A negative number indicates the existence of a capability gap.
- ***Nominal load not served*** is a performance metric derived by subtracting the power supplied from the critical demand. A negative number indicates the existence of a capability gap.
- ***Time to restore critical functions*** is both a capability and requirement metric. As a requirement, it describes the amount of time over which the inability to perform a function will not critically impact mission performance. As a capability, it is the amount of time it takes the provider system to restore service to a point during a disruption. This may be the time it takes for backup power generators to turn on during a power outage, for example. Performance is determined simply by subtraction: If time to restore power is greater than the requirement, mission assurance is lost, and there is a gap.
- ***Time to restore nominal operations*** is both a capability and requirement metric. As a requirement, it describes the amount of time over which a function can operate in a degraded state (e.g., under minimum backup power) before the stress on the system becomes so great as to compromise mission assurance. As a capability, it is the amount of time it takes the provider system to restore nominal operations. Consider a case in which a mission function relocates from one base to another during a disruption that disables a building at the home base. The secondary base can support operations for some period, but performance eventually degrades to the point of putting the mission at risk because the secondary base cannot fully support the function for very long. Performance, then, is the gap between the required and actual restoration time for nominal operations.
- ***Total harmonic distortion*** and ***voltage sags and swells*** are metrics describing power quality characteristics required and provided. Performance is determined by measuring the differences between the power quality required and the power quality provided.
- ***Outage cost*** is a tracking metric that describes all the incremental costs to the Air Force of operating through a disruption. Many system architectures might achieve the same level of energy assurance, but some of these may be prohibitively costly to implement (maintaining multiple fully manned backup locations for mission functions, for example). Understanding the costs incurred during disruptions is an important piece of determining whether to take mitigating actions or to accept risk. There are several drivers of outage cost, including the types and duration of the outage, maintenance and other manpower needed during the outage to ensure that mission-essential functions continue to operate, and the types of physical infrastructure and equipment that constitute a particular installation's system architecture. We do not prescribe exactly what constitutes "outage cost" and rather encourage users of the presented framework to define relevant cost elements and to track them during power grid disruptions.

Connecting Energy Assurance Attributes to Metrics

At the beginning of this chapter, we discussed three key elements of energy assurance—reliability, resilience, and robustness. Generally, all three concepts have to do with attributes of services provided—or of capabilities. Performance in each category can be assessed using all or a subset of the proposed performance metrics. Reliability can be assessed by looking at the frequency of supplying power of a particular quality and level (i.e., critical load not served, nominal load not served, total harmonic distortion, and voltage sags and swells). Resilience can be assessed by looking at the level of reliability when facing a disruption and, in the case of degraded performance, the time to restore critical functions and nominal operations. Robustness can be assessed by looking at system performance in reliability and resilience across a wide range of possible scenarios.

It is important to note that the same metric could be viewed as both a reliability metric and a resilience metric, depending on the perspective of the user. For example, reliability is used to characterize the actual power coming into any given point in the system, while resilience describes the means of filling any gaps between the power provided and the power consumer's requirement in response to a disruption in supply. Consider the following situation: Grid power is out, and a civil engineer–maintained backup power generator supplied the necessary power to a mission-essential function. The owner of the mission-essential function might use the "power supplied" metric to measure the *reliability* with which civil engineering provides power, while civil engineering might use the same metric to measure the *resilience* provided by the backup generator in the face of a grid outage. See Chapter 3 for a more detailed depiction of the interfaces between energy providers and users and the different ways in which they might think about resilience and reliability.

We understand that, for some bases and missions, the presented metrics will prove insufficient to fully describe the electricity-related requirements and capabilities associated with performing mission functions. For these cases, other metrics will need to be developed. In the next section, we describe an approach for selecting useful metrics, the same approach that we used to select the ones listed in Table 2.1.

Metric Selection

We surveyed the literature for attributes of "good" metrics that we used to guide our selection process.[6] These attributes are meant to provide a way for users of the RAND framework to systematically select new metrics. Different attributes are important for different types of metrics. For instance, requirement metrics might be assessed for their ability to set

---

[6] See Bernardo et al., 2012; Savitz et al., 2015; Watson et al., 2014; Willis and Loa, 2015; and Young et al., 2014. The selected attributes relate to the commonly used "SMART" criteria (specific, measurable, achievable, relevant, and time bound) for performance indicators and objectives (see, e.g., McNerney et al., 2016). The metric attributes proposed in this work retain the essence of the SMART criteria while providing an actionable set that is especially relevant in the Air Force energy assurance context.

policy goals, while capability metrics might not—they just need to be understandable to energy users who rely on requirements being met to carry out mission functions. Starting with attributes described in the literature and favoring the attributes that are well suited for direct evaluation (e.g., whether a metric is "useful in system planning" is easier to assess than whether it is simply "useful"), we arrived at five key attributes against which to evaluate potential energy assurance metrics. These attributes—validity, policy relevance, maturity, operational usefulness, and resource intensiveness—are described in detail in the following subsections.

Validity

The validity criterion measures the extent to which the metric captures the concept being assessed (Young et al., 2014; Savitz et al., 2015). In the case of energy system evaluation, the validity of a metric is related to its ability to capture one (or more) of the key elements of energy system performance, including level of service, frequency of outage, timing, power quality, and cost. Each metric being considered will align with one or more of these elements and can be said to fully meet the validity criterion if the metric provides a direct measure of the key element. If the metric is a proxy measure, it may be assessed to partially meet this attribute, and any metrics that are only indirectly related would be assessed to not meet the validity attribute.

Policy Relevance

This attribute measures the ability of a metric to be used for setting actionable policy goals. The policy relevance here can apply to the traditional view of regulatory policy but also to the internal policy of any organization. Within the area of policy relevance, we identify two subattributes:

- ***Explainable to stakeholders:*** The metric should be easily explainable and understandable to all stakeholders involved. This attribute applies for all types of metrics and can be tested qualitatively through interactions with all stakeholders involved. As a general rule, metrics defined using simple and commonly measured parameters will tend to be favored over metrics that require more-detailed or specific knowledge. An example metric that would not score well in this area might be "five nines availability." For this metric, the definition of the system boundary, detailed knowledge of probabilistic failure rates of all subsystems involved, and the historical observations of the system must all be understood to fully explain the metric. A simpler metric related to time, such as the time to restore nominal operations, may well be preferred.
- ***Targets can be set directly:*** The form of a metric should lend itself to setting targets directly, where targets can be viewed as the goals of any stakeholder related to the metric. For receivers of power, these targets will be expressed as requirements. For power providers, these targets will be relevant when designing the power system or incorporating new equipment.

Maturity

The maturity of a metric is based on the ability to systematically collect information on the metric. Identifying mature metrics allows assessment of energy systems without requiring an inappropriate level of resources. In general, metrics that satisfy the maturity attribute will be well documented in literature or based on fundamental parameters of the system itself. We recommend this attribute be evaluated both for the Air Force and for best practices in the commercial sector. Such an assessment may provide valuable insight into metrics that may be mature in civilian energy systems but would require greater resources to implement within the Air Force enterprise.

The ability to systematically collect information and data related to a metric is important for both monitoring and evaluating the state of the energy system. At a high level, the ability to systematically collect all relevant information is assessed by the degree of difficulty and level of resources required. This can be tested qualitatively by a thorough review of technologies and practices in place that are related to the measurement of the associated metric parameters. However, if the metric parameter data are not already collected directly, the assessment can be tested quantitatively by determining the costs and other resources required to implement the collection of all relevant information. For instance, if the scale of an energy-based metric, such as nominal demand, was such that data were required monthly or annually, energy system billing data could be used. However, finer resolution of the nominal demand data, such as by the week, day, or hour, may require investing in smart meters and may increase personnel workloads. When evaluating this attribute, it is important to keep in mind the scale, both geographic and time, for the potential metric.

Additionally, the ease of measuring all associated parameters is also an important consideration in the ability to systematically collect metric information. This aspect of maturity can be assessed through the decomposition of a metric into measurable parameters and a review of current energy system data collected. It should be noted that the ease of measuring these parameters may be very different when evaluating a capability rather than a requirement. Using time to restore critical functions as an example, on the requirement side, that time can be easily assessed for the mission owner. However, the actual capability delivered will depend on the detailed workings of the scenario and energy architecture, which will make the assessment of the parameter from a capability standpoint much more difficult.

Operational Usefulness

Metrics should be useful to planners (civil engineers and other energy planners on a base) and to operators (mission owners). Because various stakeholders use these energy systems at different time and geographic resolutions, scalability is also an important characteristic of a metric's usefulness:

- ***Useful in systems planning and real time operations:*** A metric is likely to be directly useful in system planning and real-time operations if it is based on the system behaviors of interest. This attribute of the metric can be assessed qualitatively by interviewing stakeholders about their use of the metric and its associated parameters in planning and operations. In general, stakeholders should be able to readily assess the usefulness of a metric.

- ***Scalable in time and geography***: In addition to a single stakeholder being able to use the metric directly, the metric should be easily scalable with respect to time and geography. The primary motivation for this attribute is the fact that metrics will be shared across stakeholders at a base, and the required scale will change from stakeholder to stakeholder. Assessment of this attribute is based on whether the associated metric parameters can be accounted for across time scales and geographic boundaries. From a qualitative perspective, interviews with stakeholders can help assess the usefulness of a metric across the time and geographic resolutions.

Resource Intensiveness

The resource intensiveness of a metric provides a general measure of the resources required for measurement. It includes consideration of manpower, equipment, infrastructure, and overall costs. As an example, the restoration time as a requirement metric would have very low resource intensiveness, since getting information would be a matter of asking stakeholders for their requirements. However, from a capability metric standpoint the restoration time could require assessing a high level of resources under outage scenarios because detailed modeling and simulation or physical testing would likely be required to collect information about the capability.

Appendix A lists the candidate metrics we considered; a scheme for scoring metrics on key attributes; and an overview of how the selected metrics, and some that we rejected, scored on the attributes described here.

3. Current Air Force Installation Power System Architectures

This chapter describes, in general terms, the existing electric power system architecture at most Air Force installations in CONUS. This architecture includes not only physical infrastructure but also relevant personnel, authorities, and data. A clear understanding of roles and responsibilities is needed to ensure successful implementation of the energy assurance framework presented in Chapter 5. In addition to civil engineers and mission owners on bases, many organizations throughout the Air Force play a role in managing energy, from the base level up to the Headquarters Air Force (HAF) and Secretary of the Air Force (SAF). An understanding of the existing architecture serves as a baseline and helps ground the discussion of the framework.

In the following sections, we describe the general electric power physical infrastructure and the key organizations and stakeholders involved in assessing and improving energy assurance on Air Force installations, starting with electricity users.

Physical Infrastructure

Generally, CONUS bases have similar electric power physical infrastructures. Figure 3.1 is a conceptual diagram of a CONUS base power infrastructure. External utility power comes to the base via one or a few main lines and through one or two substations. Typically, utility-owned substations are outside the base fence, but these substations are occasionally on base property. Bases with more-advanced, complex physical infrastructures might also have an Air Force–owned substation or switching station on base property. The configuration of electrical line infrastructure on base is unique to each base, but generally, one main commercial line enters each facility.[7] For facilities with critical assets, a separate line within the facility connects these assets to back up power generation.

When utility power goes out, resilience is mainly provided by backup generators that are primarily maintained by civil engineers. Typically, backup generators are installed and connected to provide power to only one facility. However, recent changes to Air Force guidance may allow generator support to multiple facilities or assets in the future.[8] Temporary mission

[7] Variations on electrical line configuration are driven by mission requirements. For example, a mission critical system with a requirement for redundant commercial feeds will have two separate electrical lines entering the facility.

[8] Generator installation guidance can be found in Air Force Instruction (AFI) 32-1062, 2015. AFI 32-1062 replaced AFI 32-1063, which used to contain the following guidance on backup generator installation: "A generator installed to support a mission-critical facility shall be installed and connected to only provide power to that specific facility. Utilizing one generator to support multiple facilities is not authorized because if the generator were to fail it would jeopardize multiple missions."

Figure 3.1. Conceptual Diagram of CONUS Base Electric Power Physical Infrastructure

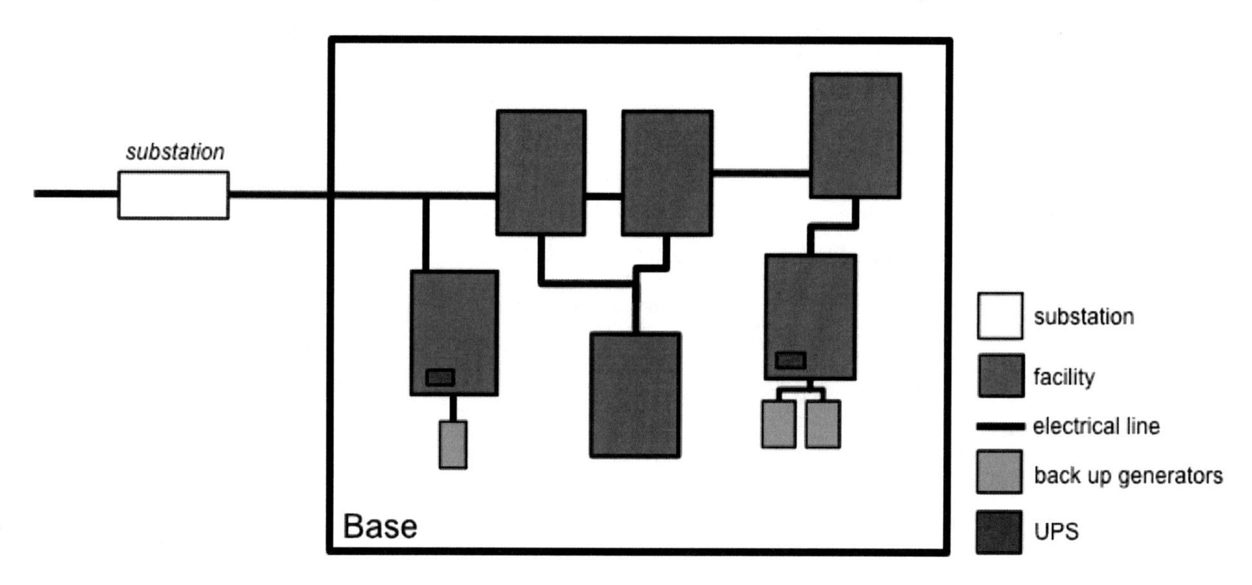

resilience is provided by UPS systems that are primarily the responsibility of mission owners and are maintained by off-base contractors.

Electricity Users

Energy users consist primarily of mission owners and the support organizations on base, such as the mission support group (MSG) and the medical group (MDG). Mission owners are commanders of organizations responsible for some kind of operational outcome. In this context, a mission could be fighter pilot training; operating and maintaining unmanned aerial systems; or operating a radar installation, command headquarters, or tenant unit for another service or government agency. The key is that the mission can be articulated, has a commander, has quantifiable energy demands, and is differentiated from other missions and the rest of the installation by some sort of organizational boundary.

Mission owners have many and varied chains of command. For example, at Davis-Monthan AFB in Tucson, Arizona, the host unit is the 355th Fighter Wing, part of Air Combat Command (ACC). Thus, the wing commander, who is also the installation commander (IC), receives direction and mission prioritization from the commander of ACC. At the same time, Davis-Monthan houses Air Force Materiel Command's (AFMC) 309th Aerospace Maintenance and Regeneration Group, the aircraft boneyard for all excess military and government aircraft. So, the commander of this group receives direction and mission prioritization information from the commander of AFMC.

Davis-Monthan has other tenants: 563rd Rescue Group and 55th Electronics Combat Group under ACC; Headquarters, 12th Air Force, including Air Forces Southern's (AFSOUTH's)

Combined Air and Space Operations Center;[9] the 943rd Rescue Group, under Air Force Reserve Command; and the 214th Reconnaissance Group, under the Air National Guard. Thus, Davis-Monthan hosts mission owners from four different major commands (MAJCOMs) (only one of which is the IC) and a warfighting headquarters.

Besides mission owner energy demands, the installation itself has significant energy demands. The MSG encompasses most of these, including civil engineering, security, communications, and a host of other activities that keep the base functioning. Medical operations are usually commanded under an MDG.

All the missions on a base and all the installation support functions compete for energy supplies. At Davis-Monthan, commander of ACC has significant influence over energy resilience decisions, commanding the host unit (and thus the IC, the 355th Wing commander) and two tenant units, but not complete control; three other MAJCOM commanders have missions located there, as well as the Combined Air and Space Operations Center for U.S. Southern Command's air component, AFSOUTH.

Electricity Providers

Mission owners and base support organizations use facilities and other infrastructure to fulfill their missions. Energy users pass on their requirements to civil engineers on the base (operating under the leadership of the base civil engineer), the energy manager, and/or other energy planners on the base. Ideally mission owners and base support organizations quantify and articulate their energy demands to the base civil engineer and other base-level energy planners using easily understandable metrics. The base civil engineer is then responsible for supplying all the day-to-day energy needs of the base and quantifies and articulates the base's needs to the local utility in the form of requirements.[10] The base civil engineer's specific responsibilities include operating and maintaining all Real Property Installed Equipment (RPIE) electric power systems and equipment;[11] testing generators in a specified way following a particular schedule; providing "24-hour maintenance support for fuels facilities and associated equipment"; and, with approval from the Air Force Civil Engineering Center (AFCEC), acquiring new generators or replacing existing ones as needed (AFI 32-1062, 2015; Department of the Air Force, 2014; AFI 23-201, 2014).

---

[9] Unlike the 12th Air Force commander, who reports to the commander of ACC and has only a training mission, the commander of AFSOUTH reports to the commander of U.S. Southern Command and has a warfighting responsibility.

[10] Installations could have privatized electrical systems, with the system owner responsible for the day-to-day energy needs.

[11] RPIE includes "those items of government-owned or leased accessory equipment, apparatus and fixtures that is [sic] essential to the function of the facility" (AFI 32-9005).

The local utility provides energy service with some reliability, and civil engineers under the leadership of the base civil engineer, along with support from mission owners as needed, takes resilience measures to make up for any gaps in the reliability of the power the utility provides. Mission owners have two key energy resilience levers to make up for any gaps in the reliability of power reaching their missions: (1) UPS systems, which provide temporary power to critical loads in the event of a grid failure, and (2) COOPs (Department of the Air Force, 2014).

AFI 10-208, 2013, p. 27, defines a COOP as an

> internal effort within individual components of the Executive, Legislative, and Judicial Branches of Government assuring the capability exists to continue uninterrupted essential component functions across a wide range of potential emergencies, including local or regional natural disasters, health-related emergencies, accidents, and technological and/or attack-related emergencies.

Mission owners, following from COOP plans specific to their missions, can choose to move a mission or a portion of the mission in the event of a grid failure, assuming they have the capability to do so.

Figure 3.2 shows the relationships between energy users and providers. Demanders—mission owners and base support organizations—appear at bottom left. The IC is broadly responsible for the installation, but the base civil engineer is actually responsible for receiving, interpreting, and integrating energy requirements and providing capabilities to the users. The utility is depicted on the right side of the figure, receiving requirements and providing energy capability for the entire base.

Reliability is defined upstream of any given point in the system, and resilience is defined downstream of that point. In this way, reliability is used to characterize the actual power coming into any given point in the system, while resilience provides the means of filling any gaps

Figure 3.2. Base-Level Organizations That Influence Day-to-Day Energy Resilience Actions

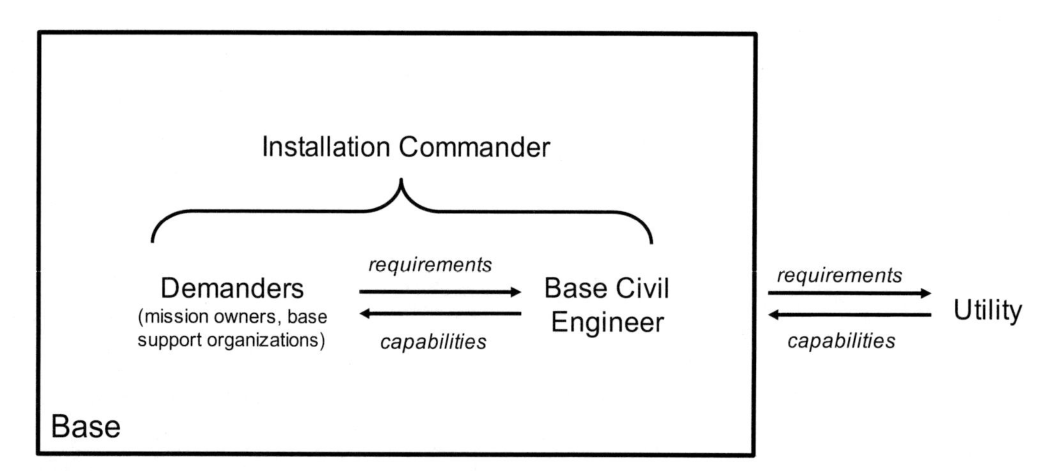

between the power provided and the power consumer's requirement. Figure 3.3 shows the flow of notional power delivery through the system and identifies points in the system where resilience measures might need to make up for insufficient supply characteristics. Power sufficiency, shown schematically in the vertical axis of the figure, represents not simply the amount of power in kilowatts available through the power system but also any other characteristics that are expected to be provided reliably, such as consistent power quality free from sags or spikes.

In Figure 3.3, electric power moves from providers at the left to users at the right, passing through the control of three major stakeholders: the commercial utility, the base civil engineer, and mission function owners (operators), who are the ultimate consumers of base power. The users provide both a nominal requirement, representing the power characteristics needed to perform their functions under normal conditions, and a critical requirement, representing the minimum power characteristics that will allow them to operate for some time through a

Figure 3.3. Electricity Stakeholder Interfaces

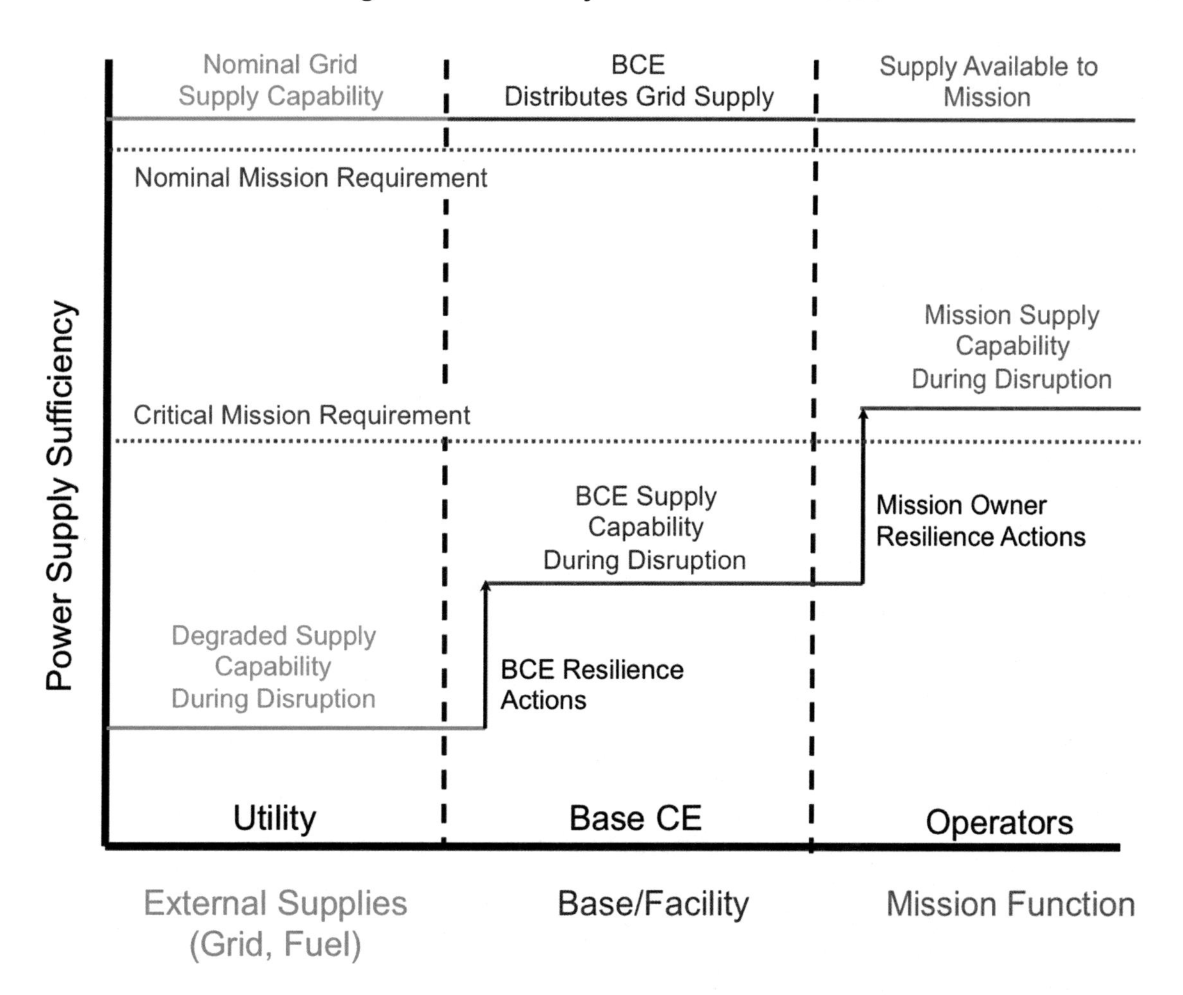

disruption. These are represented in the figure by the dotted blue lines. The critical requirement will always be at or below the nominal requirement.

Under normal conditions, commercial power is capable of satisfying all the requirements of the base, as represented by the line at the very top of the figure. Under these conditions, civil engineering personnel are not undertaking any resilience actions or providing additional power; they are simply maintaining on-base transmission and distribution systems so that the utility-provided power can reliably reach the mission owners who use it without any degradation. Similarly, the operators are simply performing their missions using the sufficient power provided.

During a disruption, the utility supply may be degraded (say, in a brownout), or completely removed (blackout). The green line at the lower left of the figure represents a notional disruption in which the power input to the base is degraded to the point that it is no longer sufficient to meet even the critical mission requirements.[12] In such a case, the base civil engineer will need to implement resilience measures (turning on backup generators, turning off power to noncritical base functions, etc.) to make up for the insufficiency. These actions are represented by the black arrow in the middle of the figure, where, in this example, the base civil engineer resilience actions improve the state of affairs, but the power provided to the operators is still insufficient to meet critical requirements. Thus, the operators must, in turn, exercise any available resilience measures. This may involve operating on UPS power, which can provide sufficiency but only for a limited time. These actions are shown by the black arrow at the right of the figure. Mission owner resilience activities could include transferring the mission to another base, which would reduce the local requirement for electricity (although it will increase demand at the other location). In the figure, such activities are factored into the gap between nominal and critical mission requirements.

An important takeaway from this discussion is that, while some groups on a base are more often users of energy than providers of energy, the distinction between "users" and "providers" is not always clear cut. This classification depends on where in the system the observation is being made. For instance, mission owners receive power through distribution lines maintained by the civil engineering squadron, carrying power produced by a commercial utility. In this example, both the commercial utility and the base civil engineer are "providers" from the vantage point of the mission owner. Similarly, in the event of a disruption in service to the power grid, civil engineering–owned and -operated RPIE generators provide power to critical mission functions. But civil engineering could be seen as an energy "user" from the perspective of the commercial power provider.

---

[12] Access to fuel supplies may also be hindered during a disruption. For example, a natural disaster that knocks out the utility power supply might also physically damage roads, hindering access to external fuel supplies. This, in turn, could hinder the base civil engineer and mission owner resilience actions that depend on fuel. The green external supply line is meant to drive everything that happens to the right of it in the figure.

Decisionmakers

A number of entities are involved in decisionmaking about energy needs and provision at any given base. The base civil engineer (having received all user requirements) provides the initial input for prioritizing investments related to base infrastructure assets. Another important base-level entity is the facilities board (FB). The FB comprises the IC (who acts as the chair), major group and tenant organization commanders (e.g., MSG, MDG), the base civil engineer, and others (as determined by the FB chair) and has broad oversight over base facility decisions (AFI 32-10142, 2013). The FB acts not in a tactical role but in a more operational or strategic role, approving plans and investments that go up to MAJCOMs or enterprise-level organizations for approval.

The FB's responsibilities include approving the project list for investments, facility space utilization, and vetting and approving a range of official documents and plans that serve the needs of the base itself or higher-level program needs, e.g., the base's investment strategy, the Air Force's Critical Infrastructure Program (Donley, 2012), Defense Critical Infrastructure (DoD Directive 3020.40, 2010), and more.

Figure 3.4 begins with the original base-level depiction in Figure 3.2 and broadens the picture to include higher-level organizations and adds directional arrows depicting the flow of policy guidance and formal inputs into the Air Force project prioritization process.

Figure 3.4. Other Organizations That Influence Energy Resilience

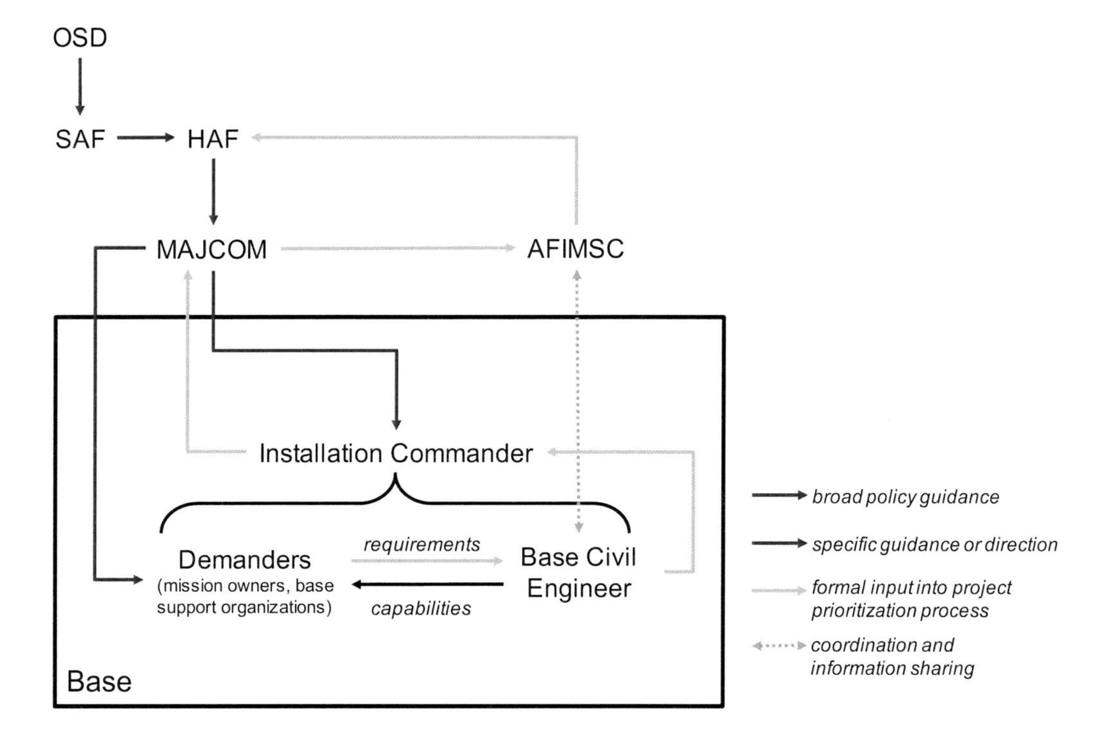

Starting at the top left of Figure 3.4, SAF is responsible for broad policy guidance and represents the Air Force to the Office of the Secretary of Defense (OSD). HAF is responsible for specific guidance and manages the Planning, Programming, Budget, and Execution process. MAJCOMs provide specific command direction to ICs and mission owners and approve and forward to the Air Force Installation and Mission Support Center (AFIMSC) prioritized project lists as inputs to the project prioritization process. AFIMSC owns the project prioritization process and directs year-of-execution funds to projects above a certain dollar threshold, once approved by HAF. Demanders, the base civil engineer, and other organizations (as directed by the IC) develop the prioritized project list for the installation as a formal input into the project prioritization process. During the development of the project list, the base civil engineer coordinates and shares information with AFIMSC, specifically AFCEC, the organization within AFIMSC responsible for compiling and validating an Air Force–wide integrated priority list.

Many other Air Force organizations have energy-related roles (e.g., virtually every major office in SAF and HAF has one or more formal positions on an energy program or committee). But our main focus as we move into a description of our framework in Chapters 4 and 5 is the base level (as indicated by the solid line around the base-level organizations in Figure 3.4): who uses energy, who controls it, and who makes decisions that affect energy assurance (and by extension, mission assurance).

Data

We surveyed five CONUS installations to get a sense of the data available relevant for the metrics described in Chapter 2. We found that most bases have a good understanding of their primary and backup sources. For instance, civil engineers at all bases had information on the kilowatt ratings of the generators on base. Similarly, all surveyed bases knew how many utility lines and substations fed the base. But none had insight into acceptable restoration times across missions or into the power-quality requirements of different missions. As to the latter, the general consensus was that critical loads that are sensitive to power-quality issues are usually connected to UPS systems that are owned, operated, and maintained by mission owners. But civil engineering has no visibility into whether this actually happens.

The degree of specificity of collected data varies across bases. For instance, at one base, energy consumption information is collected at a very granular level—monthly at the building level, or even hourly by building—whereas at most bases consumption is tracked only as an aggregated value for the whole base. Further, the format and location of energy data vary by the type of data and by base. For example, most bases track generator data in Excel, allowing quick manipulation and analysis. The same cannot be said for documentation of primary power sources and connectivity to loads on the base. These data are stored as network diagrams, line drawings, aerial maps, etc. This heterogeneity might not be a problem in itself from the perspective of any one base, but if there is an enterprise need to compare architectures across bases or to identify solutions that might work for different bases, having access to data in standard formats and stored in a central repository could help. AFCEC might be a candidate organization to integrate and manage data in a central repositorty. AFCEC already collects energy usage data from installations that are used in Air Force and DoD annual energy reports. Energy cost and consumption data at the site level are collected through the Air Force Energy Reporting System (see Air Force Pamphlet 32-10144_AFGM2016-01, 2016).

4. Scenario-Based Planning

To meet their power demands, nearly all CONUS Air Force installations depend on commercially generated electricity that is transmitted to the base. The commercial power sector in the United States has an excellent record of performance in delivering electricity at a very high level of reliability but is facing increasing challenges (North American Electric Reliability Corporation, 2015; U.S. Department of Energy, 2015, p. S-2). Outages, many triggered by extreme weather events, occur frequently enough that the Air Force—and, indeed, virtually all enterprises dependent on continuous power supplies—has invested in backup power in the event of power loss from the grid (U.S. Department of Energy, 2015, p. S-10). These outages can arise from many types of failure modes on the commercial side, such as equipment breakdowns (including cybersystems), human error, natural disasters, and determined adversaries. Outages can also occur as a consequence of or in addition to on-base events that may affect transmission lines or the various elements of on-base backup power operations.

With so many potential failure modes, mission owners and civil engineers on the base need a systematic means of thinking through the implications of plausible scenarios and identifying plans that will be robust across a wide range of possible future conditions. The purpose of this chapter is to motivate and describe the use of scenario-based planning in the context of energy assurance.

Identification of Key Assumptions

RAND and others have developed a number of approaches to short- and longer-term planning under uncertainty for military strategies, operations, and facilities over the years. Davis (2012) provides a detailed summary. One approach, known as assumption-based planning (ABP), was developed to support U.S. Army war planning (Dewar et al., 1993). ABP was based on the simple premise that plans tend to fail when critical assumptions underlying the plans are no longer valid. ABP is particularly well suited for planning environments with high levels of uncertainty that cannot be easily characterized in the form of known probability distributions. For this research, we drew on ABP for its definitions of various types of assumptions that are relevant to developing the framework discussed in Chapter 5.

ABP distinguishes among load-bearing assumptions, vulnerable assumptions, and other assumptions whose status has minor to no effect on the achievement of energy assurance as defined earlier in this report. The term *load-bearing* is borrowed from building design and refers to the assumption on which the plan's success most heavily depends. If a load-bearing assumption fails, the plan is likely to fail. As an example, suppose a base civil engineer were to assume that the off-base commercial grid always supplies sufficient electricity at nearly all times and at the desired power quality to meet all mission-essential requirements; when outages occur, the base civil engineer assumes that they will last only a few days at most. Given this assumption, the base civil engineer might choose to make minimal investments in backup power resources. However, if a major power outage did occur, perhaps as a consequence of a major ice storm or a cyberattack, the base would be unable to carry out its mission, and its energy assurance plan would fail because of failure of the load-bearing assumption of near 100-percent grid reliability. Another example of a load-bearing assumption might be that, in the event of an outage, diesel fuel to power backup generators could be procured from off base if the two- to three-day diesel supply on base were depleted.

Some assumptions are vulnerable to being undermined by future events, whether natural or man-made. Assumptions can be both load-bearing and vulnerable. Load-bearing assumptions are the assumptions on which the success of a plan or strategy most heavily rests. Vulnerable assumptions are those most likely to be overturned by future events. Load-bearing and vulnerable assumptions are the ones that are "most likely to produce nasty surprises as the plan unfolds" (Dewar, 2002, p. 3). We draw on these definitions in the next section. Assumptions are one of several dimensions of uncertainties used to develop scenarios that will be most effective in exposing vulnerabilities of bases to disruptions in electrical service that may affect mission-essential activities.

Defining Scenarios

Scenarios and *scenario planning* are often-used terms, but it is worth being clear about our definitions. Drawing from the literature, we define a *scenario* as an internally consistent story about the future, developed for the purpose of challenging the business model of an organization by thinking expansively about uncertainties (including the unthinkable) occurring in the external environment in which the organization is operating (see Chermack, Lynham, and Ruona, 2001, and van der Heijden, 1997). Scenarios are intended to deal with uncertainties of the kind that cannot easily be quantified. Planners use these scenarios to systematically assess the exposures and consequences that an organization may face, enabling it to assess courses of action available to reduce risk to its enterprise.[13]

Scenarios can take many different forms but can be defined generally as narratives describing a possible set of uncertain future conditions on which plans of any kind will depend. Schwartz (1996) is often credited with bringing the use of scenario analysis into common practice among business and government. In their most common form, scenarios are "hand-crafted" to vary future economic, demographic, technological, political, or other conditions that differ from the present and are viewed as germane to the performance of the system of interest. For example, the

---

[13] Davis (2012) provides examples of applications of scenario planning and other approaches to uncertainty analysis to national security.

Intergovernmental Panel on Climate Change (IPCC) has defined a number of greenhouse gas emission scenarios, representing different degrees of advancement of low-carbon technologies, political commitment to emission reductions, and global economic growth trajectories (see IPCC, 2014, pp. 19–26). Scenarios can also be constructed by computational means that can vary any number of uncertain factors and run a simulation model of the system of interest under hundreds, thousands, or more combinations of the uncertain factors. This approach to scenario building is at the core of Robust Decision Making (see Chermack, Lynham, and Ruona, 2001, and van der Heijden, 1997).[14]

Event-Driven Scenarios

We define *event-driven scenarios* as a set of conditions triggered by external physical occurrences, such as storms, or by the disruptive and nefarious efforts of determined adversaries. In the context of energy assurance, we assume that the Air Force has operational plans and contingencies in place for the most common scenarios. Here, we instead focus on scenarios intended to stress operational plans and assumptions to their breaking point as a way of probing the range of vulnerabilities installations may have but not know that they have.

Each event-driven scenario involves a complete loss of external power to the base, which may last for minutes or months. Some of these events can be forecast a day or two in advance, such as hurricanes and floods, although their precise consequences cannot be foreseen. The impacts of these scenarios, however, can be mitigated through response actions and shaping and hedging actions taken in advance. A challenge in responding to event-driven scenarios is the potential for a series of suboptimal actions, each measured to meet the event of the moment, to miss opportunities for more transformative change to lessen exposure to future risks. For this reason, the framework leads base personnel through all the scenarios of interest before initiating the analysis and choice of response options.

Many situations could cause a base power outage. However, in the absence of a validated simulation model of energy operations for a given base, it is impractical for that base to assess its capability to provide power to critical missions across all possible scenarios. Rather, to maintain tractability for users across the Air Force, we recommend using a small set of scenarios that span the power outage space, combined with guidance that is specific enough to enable users of the framework to generate additional scenarios of interest to them. The structure we use to describe this sample set of scenarios comprises five dimensions we call *conditions*, with a combination of values assigned to the five conditions constituting one scenario. As general guidance, scenarios should reflect key uncertainties outside the control of the Air Force, should be plausible and internally consistent, and should stretch beyond an organization's current thinking (see Chermack, Lynham, and Ruona, 2001, and van der Heijden, 1997).

---

[14] Robust Decision Making references include Lempert, Popper, and Bankes, 2003; Lempert et al., 2013; and RAND Corporation, undated b.

Table 4.1 describes the five conditions used to define an event-based scenario. Each condition is varied by extent, such as the length of time the condition holds or the magnitude of geographic area that experiences such a condition. In focusing on high-impact scenarios, we assume that the intensity of each condition is severe, such as the physical damage associated with a severe storm or the cybereffects associated with a nuanced attack by a skilled adversary. In the following subsections, we further clarify each condition.

Duration

The first condition defining a scenario is its duration. Event-based scenarios typically involve loss of the power grid, and the length of time between the start of the power grid failure and the restoration of the power grid is defined by the *duration* of the outage. The duration of the outage is unbounded; it may range from minutes to months. In focusing on high-impact scenarios, we assume that the base experiences a complete loss of power across the entire base. It is possible a base may only partially lose power. Long-term (beyond a year) reductions in the amount of available power are considered through changes in future conditions.

Physical Effects

The second condition defining a scenario describes the physical effects associated with that scenario. Physical effects may include high winds, fire, flooding, or other physical disruptions. The specific physical event is intentionally unspecified because different regions of the country are more likely to experience different types of physical effects. A base should consider both the physical effects most likely for its region and physical effects that are unlikely but possible. Different physical effects have different measures of intensity. For example, tornado intensity can be measured by the Enhanced Fujita Scale; earthquake intensity can be measured by the Richter magnitude scale; and hurricane intensity can be measured by the Saffir-Simpson

Table 4.1. Structure of Event-Driven Scenarios

Condition	Description	Extent
Duration	Period over which event causes outages or other disruptions	Hours, days, weeks, months
Physical effects	Weather or terrorist events physically damage equipment or disrupt operations	Base, local, regional
Cybereffects	Internet or information technology (IT) systems are compromised or inaccessible	Base, local, regional
Power quality effects	Voltage sags or swells that damage or otherwise degrade sensitive equipment	Present, not present
Broken assumptions and plans	Critical system architecture elements break down in unexpected ways	Backup systems fail to turn on, loss of access to base, etc.

hurricane wind scale. Associations between such scales and various levels of physical damage are well documented (see McDonald and Mehta, 2006; Pinelli et al., 2004; and Radakovich, Ferguson, and Boatwright, 2016).

How any type of physical damage affects capabilities depends on the base architecture. For example, generators on raised platforms are less likely to be damaged by flooding than generators in basements but are more likely to be damaged by strong winds. Physical effects may or may not be present; when present, their geographic extent may affect only the base (*on base*); the base and surrounding community (*local*); a larger geographic region, such as the state or several states (*regional*); or the nation as a whole (*national*).

Cybereffects

The third condition defining a scenario describes the cybereffects associated with that scenario. Cybereffects include damage to electronic networked systems through unintended or malicious interference with the ability of the system to operate as designed. The ability to remotely control parts of a base or region's electric grid can enable easy monitoring and rapid responses to problems on this grid, but such capabilities also offer the potential for remote controls to be exploited by those seeking to cause harm. The December 2015 attack on Ukraine's power grid is one example of the severe impacts a cyberattack can have on a power grid. Like physical effects, cyberattacks can have a wide variety of impacts, and bases should consider both likely and unlikely impacts. The geographic extent of cybereffects are similarly categorized, as the base (*on base*); the base and surrounding community (*local*); a larger geographic region, such as the state or several states (*regional*); or the nation as a whole (*national*).

Power Quality Effects

The fourth condition defining a scenario is whether any power quality effects are present. Voltage sags and swells can damage sensitive equipment. For example, a properly functioning UPS can help properly condition the incoming power in most cases, although extreme power quality issues may still be problematic.[15]

Broken Assumptions and Plans

The fifth condition describes effects on key assumptions embedded in response and contingency plans, not otherwise covered above, that are broken by the physical or other effects of the scenario or events external to the scenario event. These conditions could affect either capabilities or system architecture. Plans provide the "software" that governs protocols for interactions among personnel and interactions between personnel and equipment. Plans include assumptions, some explicit and some implicit, about the availability of personnel to execute certain functions when event-based scenarios occur. For example, in a severe disease outbreak in

[15] We learned of one instance in which a sudden drop in voltage, followed immediately by a sudden spike in voltage, destroyed a few servers despite UPS protection.

the region, such as influenza or Zika, key personnel in the chain of base response actions could themselves be sick when an event on-base occurs. As another example, a squirrel could chew through a newly emplaced wire connecting backup generators to critical equipment (see Mooallem, 2013). The probability that the wire will fail is assumed to be very low by virtue of its recent installation, and the problem goes undetected until an event occurs. The assumption is that newly installed equipment will work as intended until its specified maintenance check.

Risks can be reduced through regular monitoring and maintenance, but some risk will always remain. To make the exercise of running through the framework the most useful, base personnel might choose to think through broken assumptions and plans that would especially stress their particular architectures. This condition provides an opportunity to identify the most important assumptions underlying any given base's system architecture to see what the implications of the assumptions' breaking might be for mission capabilities.

Practical Concerns in Event-Based Scenario Development

Different combinations of values for these five conditions could generate hundreds of sets of future conditions; even if the duration and broken assumption conditions are limited to four possible values, these five conditions would define over 500 unique scenarios. It is not practical for a base to examine how every possible scenario will affect its base architecture. Some scenarios are not particularly interesting because they are already known to the Air Force. For example, because it is already both policy and common practice to have one or more backup generators with several days of fuel supporting buildings with critical missions, outages of a few hours or even a few days are unlikely to cause significant problems for bases. Such scenarios therefore provide little new information about what situations might stress base energy systems.

Whether pathways exist to mitigate the consequences of the scenarios or make the base more resilient to their impacts is to be determined from the analysis. Our intention is to present stressing conditions as a basis for planning, not necessarily scenarios that have known or easy solutions. For these reasons, in Chapter 5, we recommend a small number of scenarios that are stressful enough to expose weaknesses in base energy systems across the five conditions.

Anticipating Changes in Longer-Term Conditions

We also consider how the particulars of energy assurance on a base could change as a consequence of future changes in government policies and practice that do not necessarily disrupt the power grid in the short term but that could affect the investment and operational strategies of the Air Force. These changes in future conditions rarely occur as suddenly and unexpectedly as an event-driven scenario. Rather, they typically develop over years, as a consequence of an emerging consensus on the need to change policy course, although policy may change more rapidly at times in response to a major crisis or catastrophic event (e.g., the Chernobyl and Fukushima disasters). Other longer-term changes in global climate conditions (see IPCC, 2014, pp. 20–26), transformation in energy technologies, or transformations in the nature of work and business practices could affect base operations.

Responses to changes in future conditions are fundamentally different from responses to event-driven scenarios. With a longer time horizon, decisionmakers have more time to investigate and invest in various response options, enabling the implementation of transformative technologies or changes in business processes that could help mitigate the negative effects of the changes in future conditions. However, the restrictions changes in future conditions impose may impact how a base is affected by, or responds to, event-driven scenarios. For example, the base's ability to use diesel generators could be limited by future changes in air-quality regulations that prohibit all but limited use of diesel fuels. We further discuss the distinction between future conditions and event-driven scenarios in Chapter 5.

5. Energy Assurance Framework

The proposed framework is a structured approach to assessing whether and what kinds of gaps exist between energy capabilities and energy requirements at a given base or other installation under a range of scenarios. If such gaps exist, the framework provides a structured way to identify viable and response options that work well across scenarios to close the gaps or otherwise mitigate risks. The framework is focused on an individual base or installation but could be used at higher levels to support integrated analysis and decisionmaking among base leadership, mission owners, and civil engineers on bases in their respective roles of making installation-level investment and operational decisions to support missions. In this chapter, we discuss the components of the framework and our recommended scenario space. For additional guidance on implementing the framework, see Appendix B.

Logic of the Framework

Equipped with an understanding of the status quo, as described in Chapter 3, and drawing on established methods of scenario-based planning to define the critical dimensions of event-based scenarios, as described in Chapter 4, we developed an initial structure for the energy assurance assessment framework presented in this chapter. We presented our initial hypotheses about the framework to base and mission personnel, honing in on the elements they considered useful and actionable—and those that were less so. We then refined this initial framework to sharpen definitions, sequencing of analyses, and relationships among the components. These changes were guided by our objectives to create an easy-to-use, logical, and replicable approach to assessing energy assurance with sufficient flexibility to apply across the entire enterprise. For example, in our conversations, we gained a better understanding of adaptations and adjustments that could be made in the face of a surprise outage, such as quickly moving to transfer critical mission capabilities to another base out of range of the outage. We settled on the presented version of the framework after establishing that it met our objectives.

The framework was designed to help answer a sequence of questions relevant to energy assurance:

1. Are there gaps between requirements and capabilities under scenario conditions? Performance metrics of the sorts described in Chapter 2 can be used to characterize gaps between requirements and capabilities.
2. If a scenario exposes one or more unresolved capability gaps, what are the associated outcomes (i.e., mission impacts and costs)?
3. What response options are available to close gaps?
4. What constitutes a robust strategy to deal with gaps across multiple scenarios and missions?

Assessing Performance for One Scenario

To answer these questions, the framework guides mission owners and civil engineers on base through a series of steps. As shown in Figure 5.1, mission owners and civil engineers on base would gather information to characterize the essential features of a base's energy system architecture, capabilities, and requirements. They would then use this information to assess potential outcomes associated with gaps between capabilities and requirements (performance) when the base or its environs are subjected to an event-driven scenario. This process would be repeated for each scenario. Appendix B contains more-detailed guidance on how to step through the information-gathering process and provides sample questions about utility providers, power-supply sources, priority supply contracts, power outage history and likelihood of future events, backup power capacity, power requirements to mission-essential assets or facilities, base capabilities to respond to event-driven scenarios described in Chapter 4, and identification of gaps between base capabilities and power requirements.

Boxes in green represent information that is independent of scenarios. The orange represents scenarios disruptive to "normal" operations. Under these "normal" operating conditions, the inventory and description of the "system architecture" directly implies or describes baseline "base capabilities." Gray boxes represent information conditional on scenarios. The yellow diamonds represent modeling and simulation efforts or some form of discussion-based exercises needed to assess capabilities and outcomes associated with degraded capabilities.

Figure 5.1. Energy Assurance Framework: Assess Performance for One Scenario

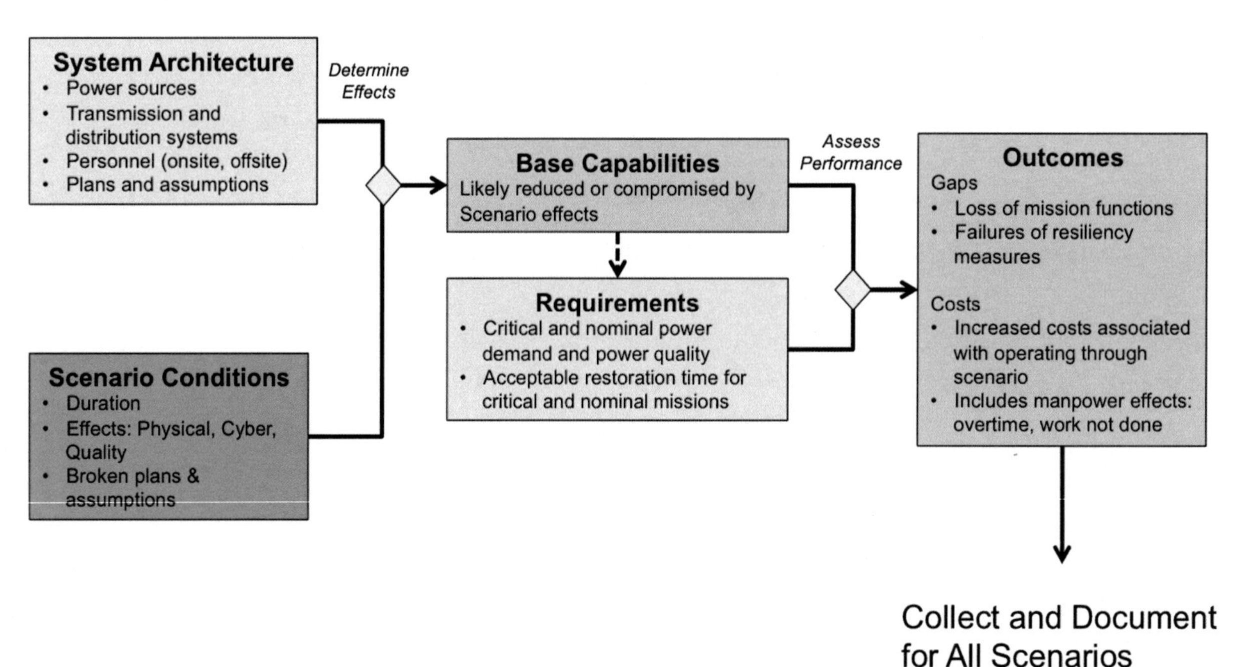

The framework, as shown in Figure 5.1, begins by identifying the essential elements of the system architecture for energy assurance, a term we define broadly as the power supply system; the transmission and distribution systems; the on-site and off-site personnel responsible for building, operating, and maintaining the power system as a whole; and the collection of plans and assumptions embedded in the plans. Characterizing system architecture is the responsibility of mission owners and civil engineers on base, who are in the best position to understand the elements and connections within their system. See Appendix B for guidance on how to identify and understand attributes of system architecture elements in a structured way.

System architecture provides the building blocks of capabilities required to perform and support missions across the base. Capabilities associated with energy assurance consist of the power characteristics provided by a supplier to a user at a given point in the system and can be described using the sorts of metrics defined in Chapter 2. As noted earlier, system architecture alone determines capabilities under normal conditions, and capabilities match requirements, although periodic review could show gaps that need to be filled to ensure mission fulfillment.

In contrast, when the base system architecture is subjected to the stress of conditions associated with an event-driven scenario, capabilities could be compromised, disrupted, degraded, or fully disabled as a consequence of the conditions associated with the scenario. Assessing the effects of scenario conditions on capabilities is a two-step process that involves (1) understanding the effects of the scenario on different system architecture elements through tabletop exercises, physical testing, or modeling and simulation and (2) calculating capability metrics (such as "power supplied") for the altered system architecture state that has resulted from exposure to a scenario. Both steps require some form of exercise or testing to complete.

Performance, then, is the gap between requirements and capabilities. *Outcomes* are the mission-specific or installation-wide consequences of gaps identified between requirements and capabilities, as conditioned on the scenario. Examples of outcomes for a particular base include loss of mission functions, needing to move the mission to another location, and inadvertent loss of lives or property. Projected outcomes could be assessed through testing, modeling, or a gaming exercise.

The framework then requires the user to document, for each scenario, performance metrics that describe the gaps between requirements and capabilities and the outcomes in terms of mission fulfillment. Also needed at this stage are estimates of increased costs incurred with operating through the scenario, whether by shifting the mission to another base or emergency procurement of maintenance personnel (e.g., overtime, displacement of other duties, new equipment, and fuel supplies).

Assessing Performance Across Scenarios

In the next steps of the framework, shown in Figure 5.2, mission owners and civil engineers on base would identify available response options across scenario outcomes; analyze these potential solution options; implement an option that performs well across scenarios; and

Figure 5.2. Energy Assurance Framework: Assess Performance Across Scenarios.

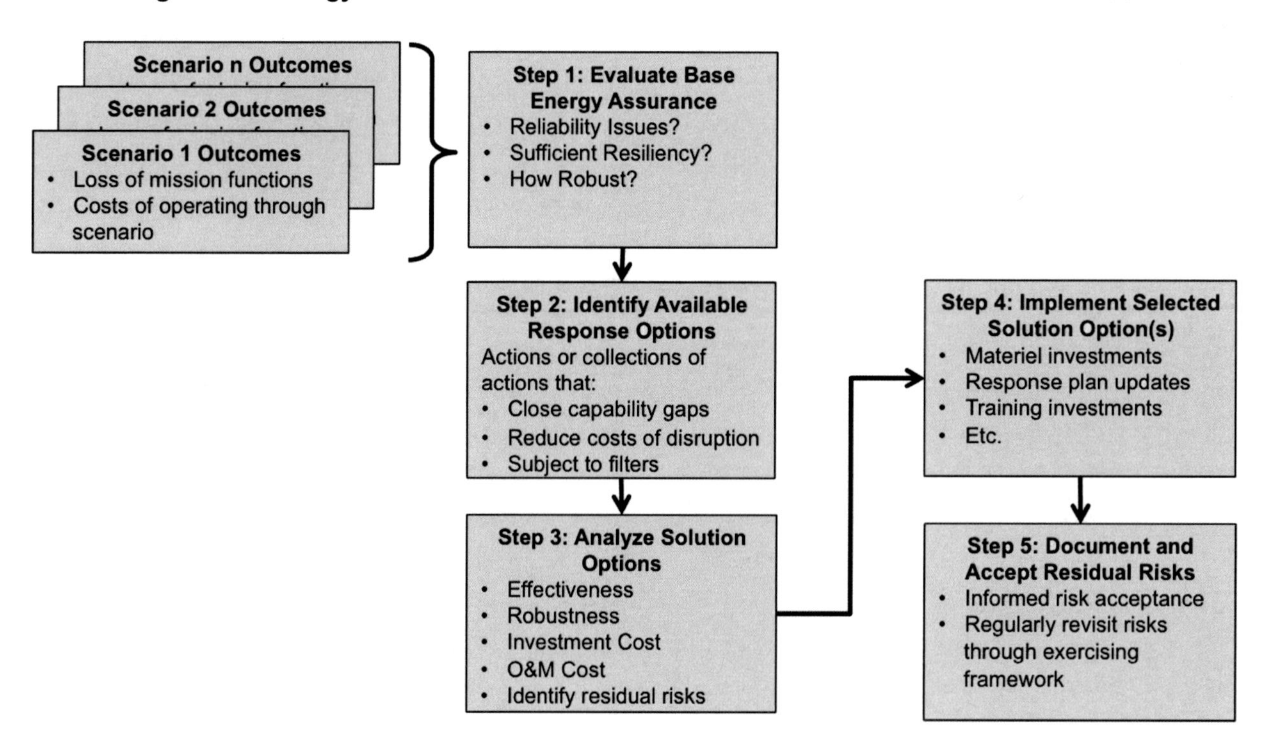

document, accept, and periodically review risk mitigation and acceptance decisions. See Appendix C for a detailed discussion of response options. The framework depends critically on applying consistent definitions of the terms and metrics described in Chapter 2. In this chapter, we focus on the descriptions of the framework's essential components and their connections to one another.

Selecting responses with the intention of optimizing performance under each single scenario is likely to be an ineffective and costly way to build resilience. Instead, the framework leads mission owners and civil engineers on base through a five-step analysis, as shown in Figure 5.2. In the first step, performance metrics are estimated and then categorized based on how they affect the foundational energy assurance goals of reliability, resilience, and robustness. This step attempts to answer the following questions: What classes of gaps exist? What seems to drive these gaps?

In the second step, a full range of response options is identified by their effectiveness in improving performance and reducing costs of disruption across the scenarios under consideration. In this step, we add the overlay of three distinct filters that have the effect of constraining the number of options for consideration. We discuss, generally, the application of these filters later in this chapter. See Appendix C for a proposed approach, using two of these filters, to reducing the larger possible set of response options to a manageable and feasible set for consideration.

The third step is to proceed with the analysis of solution options by rating them against a set of desired attributes, including effectiveness in reducing the risks associated with capability gaps and associated mission outcomes; robustness; investment costs; operations and maintenance (O&M) costs; life-cycle costs; and residual risks. Ultimately, picking the appropriate solution for a particular base will require analyses of base-specific benefits and costs (and by extension, cost-effectiveness)—similar to those REDI is conducting.

The fourth step in this segment of the framework is to implement the selected solution option through appropriate investments in materiel and training and updates to the relevant operational and contingency plans. Also included in this step would be coordination within the Air Force and, for example, with surrounding communities, regional electricity transmission organizations, and national-level organizations, as needed.

The last step in the framework is to fully document risk acceptance that accompanies the selection and implementation of the preferred solution. This step also leads back to the first step in the framework: Periodic review of these risk acceptance decisions will be critical to maintaining energy assurance at every base.

Framework Elements

The following subsections look at each of the elements of the framework in more detail, starting with the elements in Figure 5.1 (system architecture, scenarios, requirements and capabilities, and outcomes) and going on to those in Figure 5.2 (evaluate base energy assurance, identify and analyze solution options, and risk buy-down and acceptance).

System Architecture

System architecture encompasses the physical infrastructure, equipment, personnel, and planning necessary to produce the capabilities for mission fulfillment. Table 5.1 summarizes the information required to characterize system architecture and the typical owner or source of input. Assigning this responsibility to an individual or office is a prerequisite to conducting the subsequent gap and scenario analyses. See Appendix B for guidance on how to identify and understand attributes of system architecture elements in a structured way.

Scenarios

As a scenario unfolds, the base system architecture is subjected to the stress of conditions that could compromise, disrupt, degrade, or fully disable base capabilities. To keep the problem tractable and provide a concrete starting point for base personnel and leadership, we propose to focus on a small number of representative scenarios rather than consider a more exhaustive set. For purposes of scenario development, we identified a concise set of condition variables that can be used to describe a variety of event-based scenarios.[16] These scenarios are deliberately aimed at exposing conditions that base personnel are least likely to be prepared to handle and whose

---

[16] For further details on this methodology and others used for planning under uncertainty, see Davis, 2012.

Table 5.1. Inputs to System Architecture

System Architecture		Base Civil Engineer	Mission Owner
Materiel	Primary power (underground and overhead lines, points of connection to main grid, substations, incoming transmission line capacity)	X	
	Backup power sources (connectivity to mission critical loads, number, capacity, fuel type, fuel consumption, testing status, physical locations, start mode (manual or automatic), dependence on IT/internet systems for operation)	X	X
	Power conditioning equipment (connectivity to sensitive loads, number, runtime, testing status, physical locations, dependence on IT/internet systems for operation)	X	X
	Fuel (amount stored at site of backup generation, amount stored elsewhere on the base, number of storage points and physical locations, method and frequency of replenishment)	X	X
	Parts for owned equipment and infrastructure (onsite or offsite?)	X	X
Personnel	Personnel to maintain or fix owned equipment and infrastructure (onsite or offsite?)	X	X
	Personnel for emergency operations (cross-trained personnel)	X	X
Plans, policy	Communication and action plans within base and with utility or external providers during outages	X	X
	Contracts or priority agreements with utility	X	
	Mission transferability and associated resource intensiveness (money, time, effort)		X

impacts have the potential to disrupt capabilities and mission assurance. In the parlance of ABP, these scenarios might expose implicit assumptions in base plans and contingencies that turn out to be vulnerable and load-bearing. We were guided by observations and feedback on our site visits.

Example Event-Based Scenarios

Based on a literature review and interactions with subject-matter experts and stakeholders (see Chermack, 2011; Davis, 2012), Table 5.2 describes five scenarios we recommend that bases use to assess vulnerabilities and the resilience of their energy systems. These scenarios are intended to precipitate outages that could affect mission-essential capabilities over the short and longer terms. These scenarios are also intended to cover as much of the (plausible) uncertainty space as possible. Figure 5.3 presents this coverage visually, with the five selected event-driven scenarios from Table 5.2 labeled as E1 through E5.

Table 5.2. Proposed Initial Set of Event-Driven Scenarios

	Scenario E1 Delta	Scenario E2 Joplin	Scenario E3 Icestorm/Sandy	Scenario E4 Cyberattack	Scenario E5 Sandy + Cyber
Duration	12 hours	3–7 days	2 weeks	1 month	3 months
Physical effects	Base	Local	Regional	None	Regional
Cybereffects	None	None	None	Base	Regional
Power quality	Present	Not Present	Not Present	Present	Present
Scenario narrative	A lightning strike on base power line causes local fire and power quality event	High winds create large debris field on base and in surrounding community	An ice storm severely damages power lines and trips relays or a hurricane causes severe flooding and wind damage; off-base communications, landlines down	An adversary attacks IT and back-up power systems on the base and also physically targets critical nodes in the power system, cutting the power grid	Combination of Scenarios E3 and E4, where an adversary launches a targeted cyberattack following or in the midst of a Sandy-like disaster
(Sample) Broken plans and assumptions	Instruments and other equipment cannot restart following event; data unavailable	Off-base support personnel and fuel service unavailable because of downed lines and debris; communications capabilities lost	Off-base support personnel and fuel service unavailable because of downed lines and debris; communications capabilities lost	Instruments and other equipment cannot restart following event; data unavailable; loss of all communications	Off-base support personnel and fuel service unavailable; loss of access to data; loss of communications

Figure 5.3. Diverse Coverage of Scenario Space

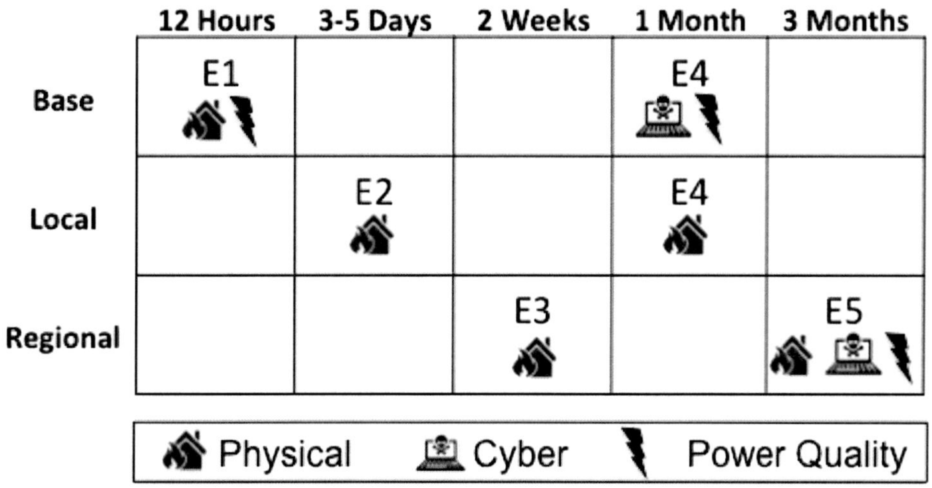

	12 Hours	3-5 Days	2 Weeks	1 Month	3 Months
Base	E1			E4	
Local		E2		E4	
Regional			E3		E5

Physical Cyber Power Quality

As shown in Table 5.2, narratives are provided with each scenario as an illustration of the situation in which the particular combination of conditions might occur. Narratives are examples only and serve to ground the discussion; many potential narratives could lead to the same combination of conditions. Base personnel should not ignore a scenario if the narrative feels irrelevant to their situation. To make this point very clear, we provide two potential narratives in the following discussion for Scenario E3, an ice storm and a hurricane. Users of the framework should construct narratives appropriate for their region.

Scenario E1, which we call the Delta scenario, involves a sudden power-quality event, followed by a short grid outage. Power-quality events can take a wide variety of forms, and bases typically use UPS equipment to control power quality for sensitive equipment, such as computer servers. Even with advanced UPS equipment, extreme power-quality events, such as several sudden drops and spikes in voltage, can still damage some sensitive equipment. In this Delta scenario, an extreme power-quality event is assumed to occur, damaging some sensitive equipment. The associated power outage is relatively short; the outage length alone should not be problematic from the perspective of providing backup power to the base. However, mission owners should pay careful attention to this scenario because this relatively likely power-quality condition carries risk and associated costs for missions, regardless of the reliability of the backup power supply.

Scenario E2, which we call the Joplin scenario, involves a grid outage of three to seven days, with physical damages on the base. For example, a tornado may have struck the base and surrounding community, severely damaging power lines and local infrastructure. Wood cross members of transmission lines, the wooden poles themselves, and even metal truss towers are susceptible to breaking or collapsing when subjected to high winds. Similarly, tornado winds can level infrastructure in their path. Bases in areas not prone to tornadoes might envision a community fire or flood. The impacts of this physical event might include a grid outage, which could potentially exceed the amount of stored diesel fuel for generators. In addition, off-base personnel could be slow to respond or entirely unavailable due to impassible transportation conditions or high demand for their services or could be busy dealing with damage to their own homes. The event could also impact communications capabilities, perhaps by damaging local cellular towers. In general, bases should ensure that response plans are established in advance to work around potential communications barriers. Bases should also consider how they will deal with fuel shortages.

Scenario E3, which we call the Ice Storm/Sandy scenario, involves a physical event that causes extreme damage not only to the base and immediately surrounding community but also to the larger surrounding region, crossing state boundaries. For example, a base might consider the North American Ice Storm of 1998, which destroyed "1,000 transmission towers, 30,000 utility poles, and enough wires and cables to stretch around the world three times" (Environment and Climate Change Canada, 2013), causing power outages across parts of Canada and New England that lasted for weeks. Bases more prone to hurricanes might consider an event similar to Hurricane Sandy, which caused electricity outages for over 8 million customers and damaged natural gas lines (U.S. Department of Energy, 2012). Personnel off base are again unlikely to be available, and communication systems may be damaged or overloaded. In addition to resolving its own issues on base, the Air Force may be called on to assist the community. Hurricane Sandy involved widespread fuel shortages (DoD, 2012c), and the Air Force, Air Force Reserve, and Air National Guard flew nearly 200 sorties moving approximately 3,000 tons of cargo (DoD, 2012b). The response to the 1998 ice storm, Operation Recuperation, involved the largest peacetime deployment of troops in Canada's history (see Bonikowsky and Block, 2016, and National Defence and the Canadian Armed Forces, 1998).

Scenario E4, which we call the Cyberattack scenario, involves a skilled adversary attacking IT systems and backup power systems on the base, in addition to physically targeting critical nodes in the power system. Depending on what critical nodes of the power system are attacked, the power grid could be unavailable for a month or longer. Whether the outage is confined to the base or includes the broader community or region would also depend on what portions of the power system are targeted. In addition to the prolonged outage, this scenario is particularly stressful for base architecture because the IT and backup systems the base would normally rely on to mitigate the impacts of the outage are at high risk. Although remote control of backup systems is normally desirable for rapid response to natural disasters, it can also present a risk for malicious attacks. The December 2015 cyberattack on Ukraine's electric grid highlights a few potential risks associated with these systems (see Industrial Control Systems Cyber Emergency Response Team, 2016, and Zetter, 2016).

The final scenario, Scenario E5, which we call the Sandy + Cyber scenario, is effectively a combination of scenarios E3 and E4. A potential malicious adversary could prepare to launch a cyberattack but wait to align timing of the attack with a severe regional natural disaster. Again, depending on what grid infrastructure is damaged, the infrastructure could take months to replace. Impassible roads, unavailable personnel, and disrupted communications systems would make it extremely difficult to maintain critical operations under this scenario. However, maintaining critical operations particularly during malicious cyberattacks is important, as adversaries may seek to create domestic disruptions to limit the Air Force's ability to counter their behavior elsewhere in the world.

Need for Periodic Review and Reassessment of Scenarios

The handful of scenarios we have presented also help outline the dimensions of the uncertainty space and help highlight the disruptive outages most likely to challenge current operational assumptions. As the dimensions of uncertainty increase, the scenarios need to become more complex. Many more scenarios could be imagined, and we encourage the Air Force and individual bases to push toward a larger and more varied set of stressing scenarios that will spur an expansive discussion of vulnerabilities and response options in the near and longer terms. We further encourage each base to review its own scenarios periodically. For example, the use of red teams could challenge the scenarios' continued utility as vulnerabilities change, as energy technologies advance, and as connections between energy and cybersecurity become more entangled. Another reason for periodic reassessment relates to longer-term changes in conditions, as discussed in Chapter 4. Future conditions could include a changed climate and a different regional economy that could alter demographics and subsequent availability of skilled personnel. Over time, future conditions may alter assumptions and system architecture. The framework itself remains focused on event-driven scenarios, and future conditions come into play during periodic reviews.

Requirements and Capabilities

Requirements and capabilities are matched sets. As previously defined, requirements consist of the power characteristics a user needs to receive from a supplier at a given point in the system. Capabilities are the means to meet requirements and are measured in the same terms. For requirements to be meaningful as a driver of capabilities, the requirements must have a direct linkage to the performance of mission-essential functions, and what is "essential" should be clearly defined. A particular unmet requirement serves no useful purpose if the failure to meet it carries no penalty. For example, a requirement for a given mission function to achieve a power quality reliability of "five nines" implies that failure to achieve that requirement leads to failure to achieve that particular mission function and, potentially, failure to achieve the mission.[17] Otherwise, five nines may be an aspirational feature, but not rise to the level of a requirement.

We selected a relatively small number of metrics to describe requirements and capabilities (see Chapter 2), some of which apply to both components of the framework. These metrics provide the basis for constructing the performance metrics, which characterize gaps between requirements and capabilities. Requirement and capability metrics cover three categories: amount of power required and supplied, quality of power required and supplied, and acceptable (requirement) and actual (capability) restoration times for different disruptions. Requirement metrics include critical demand and nominal demand, acceptable time to restore critical and nominal operations, and acceptable power quality (characterized in terms of total harmonic distortion and the presence of voltage sags and swells). Capability metrics include total power supplied, actual time to restore critical and nominal operations, and the provided power quality (characterized in terms of total harmonic distortion and the presence of voltage sags and swells). Table 5.3 summarizes the information required to characterize requirements and capabilities and the owner or source of input.

---

[17] *Five nines* means 0.99999 availability, or being up 99.999 percent of the time.

Table 5.3. Sources of Data to Characterize Requirements and Capabilities

Requirements		Base Civil Engineer	Mission Owner
Power demand	Critical loads (number, size, physical locations)	X	X
	Power requirements for critical loads (average monthly or yearly, average daily, peak, dynamics)	X	X
	Power requirements for noncritical loads (average monthly or yearly, average daily, peak, dynamics)	X	
Power quality	Sensitivity of critical and noncritical loads to power quality effects		X
Restoration time	Acceptable downtime for critical loads	X	X
	Acceptable downtime for noncritical loads	X	

NOTES: *Critical* here refers to the criticality of a power source to the functionarity of a particular mission. Higher-level Air Force leadership dictates the relative criticality of missions on a base.

Relationships Among Scenario Conditions, System Architecture, and Capabilities

Scenario conditions stress components of the system architecture and subsequently compromise capabilities to some degree. For a scenario to be useful for this analysis, mission owners and civil engineers on base will need to draw a clear connection between the conditions associated with the scenario and the disruption, diminishment, or disabling of the link between one or more elements of the system architecture and one or more elements of capabilities. Determining effects on capabilities (as expressed by the capability metrics previously defined) requires either physical testing, modeling, and simulation efforts (for example, using the tool under development by MIT Lincoln Laboratory [Judson et al., 2016] or, in the absence of a mathematical model, determining the relationships through a tabletop game or expert elicitation.)

Table 5.4 provides examples of mapping system architecture elements to a particular scenario condition and, from there, to a reduced capability. Suppose the base depends wholly on short-duration UPS and diesel-fueled backup generators, with three days of diesel fuel stored on site. A power-loss scenario that extends for a week (a shorter version of Scenario E3 in Table 5.2) could reduce the power generated by the diesel generators. How much backup power would be diminished would depend on other scenario conditions—for example, physical effects, such as the accessibility of nearby roads and functionality of off-base fuel suppliers who could replenish depleted on-base supplies, and cybereffects, which could disable some of the backup systems and advanced electronics on which critical missions depend. Thus, the duration of a power-loss scenario (scenario condition) will have a bearing on the longevity of backup power resources (system architecture) and could reduce critical mission activities because of load not served or insufficient power quality (capability).

Table 5.4. Sample Linkages Among System Architecture, Scenario Conditions, and Capabilities

System Architecture Component	Scenario Condition	Capability Component	Examples
Primary power	Duration of outage; physical effects; power-quality effects	Power demand and power quality unmet	Tornado takes down main power line to base
Backup power sources	Duration of outage; physical effects; power-quality effects	Power demand and power quality unmet	Fuel for generators cannot be delivered to base because of flooding or storm debris
Power conditioning equipment	Physical or cybereffects	Power demand and power quality unmet; unacceptable downtime for critical loads	Servers overheat and become unusable
Personnel to maintain and repair equipment and infrastructure	Physical effects	Power demand and power quality unmet; unacceptable downtime for critical loads	Off-base personnel slowed or prevented from reaching base to operationalize contingency plans
Personnel for emergency operations	Physical effects	Power demand and power quality unmet; unacceptable downtime for critical loads	Emergency responders ambushed by determined adversaries
Communications and action plans	Broken assumptions	Power demand and power quality unmet; unacceptable downtime for critical loads	New base leadership unfamiliar with details of communication protocols in emergency response plans
Contracts and priority agreements with utilities	Broken assumptions	Power demand and power quality unmet; unacceptable downtime for critical loads	Changes in utility ownership disrupt priority agreements
Mission transferability and associated resource intensiveness	Broken assumptions	Power demand and power quality unmet; unacceptable downtime for critical loads	Primary base for transfer lacks personnel because of other mission-essential activity

The linkages among system architecture, scenario conditions, and base capabilities are further moderated by the existence and functionality of monitoring and warning systems available to mission owners and civil engineers on base. The most common of these systems would be hurricane warnings, typically received as early as 72 hours in advance of arrival. Tornado warnings have a shorter lead time, typically less than a few hours. Assuming such warning systems are fully operational and not themselves vulnerable to disruption by the scenario, they can mitigate the impacts of scenario conditions on base capabilities. However, the proper functioning of these warning systems is an example of the kind of assumption embedded in most contingency plans and COOPs. Including these systems in the framework is intended to make the identification of these assumptions explicit when considering potential impacts of scenarios on capabilities.

Outcomes of an Individual Scenario

Outcomes come in two forms: (1) the loss of mission functions or failure of resiliency measures resulting from gaps between requirements and capabilities and (2) total marginal outage costs associated with operating through a scenario, including manpower effects, such as overtime or opportunity costs associated with other work not done. Outcomes could take the form of failure to achieve a mission; the unexpected need to shift a mission to another base; inadvertent loss of property; emergency procurement of new equipment; and, in the extreme, loss of life. To have a bearing on energy assurance, gaps between requirements and capabilities (performance) need to have consequences for mission assurance (outcomes). The proposed performance metrics in Chapter 2 capture three types of energy assurance gaps:

- critical and nominal load not served
- unacceptable restoration time for critical and noncritical functions
- unacceptable power quality.

Outage costs are associated with each of these types of gaps. As discussed in Chapter 2, we treat outage cost as a tracking metric that describes all the incremental costs to the Air Force of operating through a disruption. Costs are important when considering alternative investments to close energy assurance gaps. Even in the absence of energy assurance gaps, very high outage costs may drive investment in architectures that are more cost-effective. The ultimate goal of considering this cost metric is to provide monetary valuation for benefit-cost or cost-effectiveness assessments. When comparing scenarios, these costs are important when considering solutions intended to reduce unmet demand and gaps in restoration time and power quality—all of which carry their own costs. If outage costs are less than solution costs across scenarios, then maintaining the status quo of base architecture and capabilities is likely to be the preferred action. These costs could be evaluated by the base civil engineer, with input from mission owners.

Loss of mission functions and their associated costs would be tabulated for each scenario and its associated cycle through the framework. These scenario outcomes, coupled with the performance for each scenario (described in terms of performance metrics) then become inputs to the sequence of steps in Figure 5.2, in which base energy assurance is assessed across scenarios.

Evaluate Base Energy Assurance: Step 1 of Figure 5.2

After working through the framework cycle for each scenario, the next step is to synthesize base performance across the individual scenarios. This is the step of the framework that answers the question of whether the base has problems and, if so, of what types when it comes to providing the requisite level of energy assurance, subject to a diverse set of stressing scenario conditions.

A frequent trap in incremental planning is to select a response option to deal with the consequences of an individual scenario, often the last one experienced (i.e., fighting the last war). This approach can lead to brittle, ineffective, and costly investments without necessarily delivering energy assurance. Even with the limited approach to scenario analysis that we propose here for tractability, there is still the potential to improve on a one-scenario-at-a-time choice of response options. The aim is to identify response options that perform well across a full range of scenarios and can also be shown to be cost-effective in terms of reliability and resilience.

The first step in looking at performance across scenarios is to figure out which types of gaps between requirements and capabilities seem to be the most problematic for the base and to think through the drivers behind gaps. For instance, for a given base, the result of exposure to each scenario might be the same—critical load not being served. But the reasons for this end result might differ across scenarios. In a notional Scenario A, backup generators might turn on, but access to diesel fuel might run out after the second day of the outage. In a notional Scenario B, generators might be physically intact but simply fail to turn on. In a notional Scenario C, UPS systems connected to mission-essential loads might fail, damaging sensitive loads. All three scenarios would result in critical load not being served.

In this step of the framework, users should first identify commonly occurring gaps across scenarios and dig deeper to identify the underlying causes. Depending on the vantage point of the framework user, the underlying causes might seem like *reliability* issues or *resilience* issues. To the base civil engineer, who owns and operates backup generators, the issue of generators not turning on in Scenario B might look more like a *reliability* issue that the base civil engineer has introduced to the mission owner. The same issue presents itself as an opportunity to employ a *resilience* measure for a mission owner. To the utility, base generators not turning on is squarely an issue of *resilience*. The point is to understand the specific gaps between requirements and capabilities, not to first classify problems as falling into the reliability or resilience bins.

Finally, the framework is intended to identify and highlight the responsibilities of key decisionmakers and actors when it comes to addressing different types of problems. Scenario A should leave the base civil engineer thinking about generator reliability; Scenario B should induce the logistics readiness squadron or others who handle fuel supply and resupply on the base to think about ensuring access to necessary fuel during long-term outages; Scenario C should force the mission owner to pay attention to routine maintenance of UPS systems.

Identify, Analyze, and Implement Solution Options: Steps 2, 3 and 4 of Figure 5.2

The motivation behind this component of the framework is that many potential options for investment (especially in new equipment, systems, or infrastructure) are available to mission owners and civil engineers on base, and indeed, they are frequently the targets of vendors of new power and control technologies intended to solve *some* problem. Whether it is the right problem is not always clear. The steps that precede this one in the framework are intended to bring clarity to the question of whether and what types of problems exist. The following steps provide guidance on how to think through the match of responses to identified problems. Our goal is not to provide specific answers, such as "install a microgrid at Base A if x, y, and z things are true about this base." Given the wide diversity of resources and constraints across bases, such an approach would be inappropriate. Rather, we provide some guidance in this chapter and in

Appendix C for identifying the appropriate types of solutions for different problems and for considering resources and constraints in a systematic way before deciding whether to invest and what investments to make.

Solutions should be picked primarily on the basis of their ability to reduce different gaps between requirements and capabilities (i.e., their ability to improve performance) and/or their ability to reduce outage costs. Each gap the framework identifies may have more than one root cause, on the base system architecture and the scenario conditions that led to the gap. The nature of the specific problem will influence which response option or set of response options is most appropriate for addressing the problem. Appendix C provides an approach for identifying *candidate* sets of solutions from options for reducing specific gaps. We outline the broad approach here, with a more detailed discussion in Appendix C.

We identify in Appendix C possible solutions spanning the following four categories that could help reduce each of the three gaps described in Chapter 2—load not served, unacceptable restoration time, and unmet power quality requirements:

- personnel actions
- operational changes
- equipment or technology investment
- infrastructure and facility recapitalization.

Filtering of Options

The optimal response to each scenario is a function of the decisionmaking environment. For example, responses to scenarios that rely on wind or solar energy may work in some locations but not others; connecting on-base generation to local networks may enable faster responses in scenarios with physical effects but may increase risks in scenarios with cybereffects. We propose a filtering process to arrive at a manageable set for consideration. This process uses the following filters:

1. **Base-specific factors for investment decisions:** Certain attributes of base physical infrastructure or location may influence feasible investment options. For example, wind, solar, and other renewable power sources may not be available or cost-effective in all places.
2. **Scenario-related factors for execution decisions:** The decision to execute a given response during an outage may depend on the specific scenario conditions. For example, some potential responses may depend on off-site personnel, access, and services that may not be available under all scenarios.
3. **Game-changers affecting longer-term conditions:** Possible changes or shifts in future conditions could constrain or render obsolete proposed solutions. These include changing climatic conditions, technological transformations in energy generation and distribution, changes in workforce composition and availability, and other changes we cannot yet foresee. Probabilities are not associated with these potential long-term conditions, but present evidence suggests that they are plausible (i.e., possible and even likely) from the technological and policy perspectives.

Filters 1 and 2 can help identify needed investment (materiel, training, personnel, etc.), although picking the final appropriate solution for a particular base will require analyzing costs and other base-specific considerations (similar to the analyses REDI is conducting). The proposed approach in Appendix C for reducing the larger possible set of possible responses to a manageable and feasible set for consideration uses only the first two filters. However, before investing in solutions to the event-driven scenarios, decisionmakers should also consider the third filter: possible changes or shifts in future conditions that could constrain or render obsolete proposed solutions. Such shifts may occur slowly over time and are driven by changes in policy, environment, or economic conditions.

Table 5.5 presents a sampling of changes in future conditions that decisionmakers should consider. Broadly, these include operational and personnel changes that affect the ability to obtain or retain skilled personnel and changes in the ability to invest in equipment, technology, infrastructure, and facility recapitalization.

Reduced defense budgets because of sequestering, force reduction, or other changes could limit the Air Force's ability to invest in both personnel and infrastructure. Investing in solutions that require intensive personnel support or significant financial support could be unsustainable in future scenarios with reduced budgets. Alternatively, engagement in a major war might increase the Air Force budget, but a greater percentage of personnel may be deployed overseas. Increases in the tempo of force rotation could mean personnel have less experience with the base's unique systems. Changes in future personnel availability could also be due to changing economic conditions. If demand for civil engineers or energy experts increases faster than new engineers

Table 5.5. Sample Changes in Future Conditions

Solution Category	Changes in Future Conditions to Consider	Potential Drivers
Operational and personnel changes that require manpower (with specialized skills or otherwise)	• Drastically reduced military budgets • Large-scale deployment of Air Force personnel overseas • Difficulties retaining or growing the civil engineer career field	• Global shifts in defense priorities • The United States is engaged in a major war • Demand for civil engineers and energy technologists goes up, making it harder for the Air Force to recruit and retain the best and the brightest
Investments in equipment, technology, infrastructure, facility recapitalization	• Changed prime and backup power supply mix: no nuclear; renewables only (no diesel backup) • Changed grid configuration: regional or national grid with increased distributed generation sources, or increased penetration of microgrids; no national/regional grid—only distributed generation and microgrids	• U.S. nuclear plants shut down after major accident; no exemptions granted for the Air Force • Carbon legislation passed; increased penetration of renewables and other distributed generation • Public consensus to fully phase out all uses of fossil fuels and nuclear power; renewables only in power mix; maturation of power storage technologies

can be trained, their wages will increase, making it harder for the Air Force to recruit and retain the best and brightest.

Changes in the sources of electric power themselves could also affect not only base architecture but also the likelihood of various event-driven scenarios, their effects on capabilities, and the solutions available to mitigate the impacts. A Fukushima-like event could sway the United States away from nuclear power. The availability and reliability of the power grid would be sensitive to the types of power sources that replaced whatever generation capability had been removed. Similarly, further economic, environmental, or political pressure could lead to regulations that push renewable generation and limit or disallow the use of such systems as diesel generators. Any investments in disallowed systems would be sunk costs; the Air Force would need to spend additional money on new systems that are permissible under the new regulations. In addition to different power sources, the structure of the grid itself could shift from a regional or national grid to relying more on distributed generation and microgrids. Such shifts would have further implications for the reliability of the external power supply against natural disasters and determined adversaries.

Establishing Degree of Difficulty

We further categorize solution options into "tiers" based on the degree of difficulty of gaining approval, funding, and implementation:

- **Tier one** consists of actions that are allowable under existing policy and authorities, do not require new funding, and utilize existing resources.
- **Tier two** consists of actions that require financial investment outside current FY programmed funds or modest policy changes.
- **Tier three** consists of actions that require substantial investment or involve new policy or major changes to existing policy.

See Appendix C for detailed definitions of these tiers and example tier classifications of possible solutions.

Benefit-Cost Analysis

So far our discussion of solution options has focused on their effectiveness (their ability to close capability gaps identified in previous steps of the framework), their technical feasibility (location and scenario filtering), and their relative implementation ease (tiers). Air Force leadership, base commanders, mission owners, and civil engineers on base need to also prioritize options according to how much they cost and, specifically, how much they cost *relative* to what it would cost to do nothing across the scenarios considered. This is where the tracking metric, outage cost, comes into play.

One approach to ensuring that solutions are cost-effective for a diverse set of scenarios would be to conduct a standard benefit-cost analysis using a spreadsheet or simple optimization routine. The heuristic would be the same in either case. The process would start with the solutions that address the capability gap (i.e., that are already deemed to be *effective*). For each potential solution, the benefit would be the magnitude of the gap it addresses under any of the scenarios. Some options could be capable of addressing several gaps; others could be exclusive to a gap associated with a single scenario. For convenience, some of the individual responses could be assembled into sets that could be considered together. In the next step, the costs of each option would be estimated using best available information from procurement specialists. These costs could be compared to both the benefits of reducing the gaps and, consequently, of buying down the risks in terms of loss of mission capabilities—and the costs associated with the full set of scenario outcomes. If the costs of the options outweigh the costs of outages, as defined earlier, there would need to be other mitigating factors that would lead to pursuit of that particular response.

The end product of this analysis would be a set of possible responses, the risk buy-down associated with each set, and the costs associated with implementing each set. Trade-off curves could then be constructed that, for example, relate costs to residual risks. Decisionmakers would then choose where they would be most comfortable along the trade-off curve. A critical facet of this discussion about trade-offs will be the ability to differentiate between the gaps that matter more and the gaps that matter less in terms of risk to mission-essential capabilities. As noted previously, not all gaps lead to bad outcomes. Further, not all risks can be eliminated at an acceptable cost. We next elaborate on the concepts of risk buy-down and risk acceptance and how these ultimately subjective decisions can be shaped.

Risk Buy-Down and Risk Acceptance: Steps 4 and 5 of Figure 5.2

The purpose of this stage of the framework is to look across the scenarios analyzed and tabulate the risks to mission and costs of outages racked against the benefits and costs of potential solutions (or combinations of options) that are responsive to those risks. The benefits of the options will need to be expressed in terms of their reductions in risk to mission (risk buy-down) for each scenario. To the extent that an option does not fully buy down risk, a risk residual remains. Costs of outages can be compared to costs of potential solutions. For each scenario, options can be compared by their risk buy-downs and costs.

The analytical challenge is to integrate scenario-specific trade-offs across all the scenarios of interest. With a relatively small number of scenarios, a solution that does well across most or all the scenarios could be relatively easy to identify. As the number of scenarios increases, a decision support tool would be necessary to sort among the many more potential solution options to identify a robust solution. An optimization-based tool, such as those RAND developed in multiple applications for use in selecting water-related infrastructure projects, could be used for this purpose.[18]

---

[18] For examples of decision support tools applicable to energy assurance decisionmaking, see Delta Stewardship Council, 2017; SEDAPAL, 2015; and RAND Corporation, undated a.

Informed decisions will need to be made to act or do nothing and live with known risks. A decision to act requires prioritizing risks associated with scenarios and missions (for headquarters' decisionmakers); conducting detailed analyses of costs and risk buy-down for solution options; and engaging in a deliberative process of choosing solutions consistent with what is possible given availability of funding, planning horizons, and perceptions of the likelihood of a changing risk climate.

Implementation of Selected Solution Option(s): Step 4 of Figure 5.2

Once a solution option is chosen, the implementation process may depend on the "tier" in which that solution was categorized. The more challenging Tier 2 and Tier 3 solution options may take more time to implement. Such considerations will need to be kept front and center as the base civil engineer continuously monitors base vulnerabilities.

Document and Accept Residual Risks: Step 5 of Figure 5.2

Whichever solution is ultimately chosen, the questions of which risks have been addressed and which risks are likely to remain unaddressed will need to be documented from the perspectives of the civil engineers on base, mission owners, and headquarters-level Air Force leadership. Who makes these decisions will depend on the governance arrangements the Air Force chooses to apply in the context of its ongoing movement toward an enterprisewide mission-assurance paradigm. A mission risk at one base might be mitigated by investing in another base that supports the same mission. Base personnel should therefore understand the capability gaps and associated risks for their own bases while recognizing that not every gap needs to be filled at every base.

As noted above, risks can never be fully eliminated, but can be understood for their consequences and costs and either accepted or "bought down." Unaddressed gaps do not represent a failure on the part of mission owners or engineers. No system will provide optimal performance across all potential scenarios, and reaching such a goal is likely infeasible due to budgetary restrictions. However, both base leadership and mission owners should be aware of the risks that are and are not covered. If it is deemed unacceptable for a certain risk to remain unaddressed, the filters for the options should be reviewed with the mission owners and base leadership to determine whether an expanded set of options should be considered. An informed decision can then be made either to accept that risk or to consult senior Air Force leadership to identify whether changes in policy or funding levels might enable that risk to be addressed. Finally, base leadership should schedule a regular reassessment of these risks and potential solutions. Changes in technology, costs, or policy may enable new solutions to previously unresolved problems.

6. Findings and Recommendations

The focus of this project is on providing a way to assess and improve energy assurance, not actually to provide an assessment of the current state of energy assurance at Air Force installations. But some findings and recommendations did emerge from our site visits and other discussions with Air Force personnel. Additionally, we gained insight into which parts of the framework might be most useful to practitioners and identified necessary steps for implementing the presented energy assurance framework. We summarize these findings and recommendations in this chapter.

Key Findings from Discussions with Air Force Personnel and Site Visits

Understanding System Architecture

Air Force installations in CONUS generally have similar electrical system architectures. In all cases, the utility provides power with some reliability. Some bases are actively engaged in discussions (e.g., through regular meetings) with the utility provider regarding planned upgrades to the grid and other changes with potential implications for the reliability of power being supplied into the base. Commercial power, often with only one or two lines and one or two substations, feeds the entire base. In some cases, the lines and transformers are redundant, and there is enough power from one set to cover the base in the event of problems with the other. However, even in this case, only personnel from the commercial utility provider can typically switch power from one line to the other.

On-base resilience in the event of a power grid disruption is primarily provided by generators (owned, operated, and maintained by the base civil engineer) and UPS systems (typically owned and operated by mission owners, with maintenance support provided by off-base contractors[19]). COOPs ensure that the mission continues even if the base is unable to operate fully (e.g., by moving the mission to another location). Access to fuel for generators is a key component of resilience capabilities. Scenarios that combine outages with constrained fuel delivery and/or reduced access to off-base personnel and parts can severely disrupt missions.

Most civil engineering organizations are aware of single points of failure. Examples of such single points of failure include a critical communications node and an aging transmission line that serves half of a particular base. But resilience capacities are often determined within organizational silos, not developed in the context of whole-base resources. This limits opportunities to develop novel solutions and visibility into dependence on single points of

---

[19] On rare occasions, the base civil engineer might also own and operate UPS systems.

failure. For instance, when a new facility is planned for construction at a base, the base civil engineer might not be involved in discussions about what the energy footprint of that facility might be and how it might affect access to and prioritization of backup power resources on the base.

Most bases have a good sense of their generator inventory and are diligent about maintenance and testing, despite occasional mission owner resistance. We did hear of at least one case in which complete tests of the backup power system had not been performed for excessively long periods, and when such a test was done, it found that the generators were not connected to all the critical systems that were supposed to be backed up. Test results are logged and sent to AFCEC, but it is unclear whether there are any real consequences for not submitting test results regularly.

Defining and Communicating Requirements

Requirements (e.g., acceptable mission downtime) are often poorly defined or communicated by mission owners. Even when requirements are specified, it is not always clear what drives them or how they tie to mission goals. For instance, some missions specify a requirement for five nines reliability. Such a requirement is not readily actionable from an energy provider perspective, and it is not clear that there is a sound basis for setting it at that value. Power requirements may also be underestimated due to unexplored interdependencies between electricity and other mission-essential infrastructures. A key example of such an interdependency is between water (e.g., pump systems) and power, a link that could cause problems for installations and critical mission functions that rely heavily on access to water if the link is not considered in calculations of power requirements.

At times, requirements seem to be based on what can be achieved rather than on true mission goals. For instance, we found that the allowable downtime at similar facilities in different locations would vary depending purely on what each facility is able to provide and would change at any given facility if the on-site capability were degraded. Such changes are acceptable if they are accompanied by an explicit acceptance of increased risk or a shift in resources to accommodate the reduced capability, but we saw no evidence that such adjustments were being made.

We also observed cases where the failure to meet requirements appeared to have no consequences. For example, for one mission function, the size of the diesel tanks supporting the generators had recently been reduced, leaving that function unable to meet its requirement for backup operation duration, but no actions had been taken or planned to address the issue. Real requirements should have real consequences for the mission and should not be altered based on capabilities.

Mission-owner power requirements are often not communicated clearly to the base civil engineer in a timely fashion before a disruptive event has taken place. This can lead to a lack of coordination among the different stakeholders and can reduce the ability to plan in advance.

Assessing Capabilities in the Face of Power Disruptions and Coping Mechanisms

The effects on base capabilities of exposure to different power-disruption scenarios are currently not systematically assessed. Not surprisingly, base personnel are typically well aware of the consequences for base capabilities of exposure to events that have previously taken place and are ready to take necessary actions to cope. For example, one base we visited routinely deals with flooding events. Civil engineers at this base are used to taking such measures as sandbagging locations on base that are known to be vulnerable. Similarly, mission owners have invested in flood-proofing measures, such as special gates to keep rising waters away from critical infrastructure. On the other hand, there is limited understanding of if and how base capabilities might degrade when facing disruption scenarios that the base has not experienced to date.

One base we visited has a large number of diesel generators, both stationary and mobile, and of a variety of makes and models. While this diversity of assets requires more training and knowledge to manage and creates more work for base civil engineering staff, it very likely increases the base's resilience to model-specific problems or attacks. On the other hand, proposals to connect a large number of generators to a virtual local area network that enables them to be remotely controlled require careful consideration. The base civil engineering staff and mission facility staff both value such a capability because it could make routine work easier and save valuable time in an outage. While this feature is beneficial in some scenarios, it could in fact prove to be a liability. For example in a cyberattack, a sophisticated and targeted adversary could render networked generators useless. It is not clear that such trade-offs are systematically assessed before decisions are made about investments aimed at boosting resilience.

Recommendations

Our site visits and discussions with base and mission personnel confirmed the need for an analytical framework, such as the one proposed in this work. A framework that incorporates clearly and simply defined metrics; clear guidance on roles, responsibilities, and necessary communication channels; and a systematic way to think through vulnerabilities (i.e., assessing the effect of adverse scenario conditions on system architecture elements) can go a long way toward making risk-informed decisions when it comes to energy assurance.

The following steps can help mission owners and base energy planners use the RAND framework and generally better understand and articulate energy requirements and capabilities:

- Mission owners should define energy requirements up front and clearly communicate them to the base civil engineer. Currently, the prioritized asset list drives base civil engineer activities during outages, but civil engineers are often forced to react to demands in real time.
- Energy requirements should be clearly tied to mission goals and needs. If an unmet requirement seemingly has no implication for the mission, it is not a real requirement.

- Assessments of electric power requirements should account for interdependencies among electricity and other mission-essential infrastructures, such as water.
- Installation energy planners and mission owners who rely on assured access to electric power should use metrics, such as the ones proposed in Chapter 2, to articulate requirements, capabilities, any gaps between the two, and the implications of the gaps.
- Operators and planners should invest in gaining a better understanding of the effects of exposure to scenarios that have not previously been experienced. A thorough assessment would require physical testing, and modeling and simulation efforts. But simple tabletop exercises, such as the one outlined in Appendix B, that focus on asking questions that reveal implicit biases about how systems and people operate on the base can yield critical insights into the extent to which an installation is truly prepared to face disruptions of different types. Such exercises can also raise awareness of important trade-offs between increased efficiency and reduced security. For instance, some investments, such as remote monitoring of generators, increase resilience to flood scenarios but may diminish capabilities in cyberattack scenarios. As with having a variety of generator makes and models, not putting all backup systems into a single basket can increase resilience to a variety of scenarios.
- Decisionmakers should look across missions at a given base and across bases that support a particular mission before investing in energy assurance upgrades. Taking a holistic look can help ensure that requirements are not identified in isolation, that capabilities are not developed in silos, and that resources are efficiently used. For instance, a mission risk present at one base could be mitigated by investing in another base that supports the same mission and is better suited for cost-effective resilience upgrades.
- In picking solutions to implement at different bases and in addition to analyzing costs, risks, and other base-specific considerations, decisionmakers should consider possible changes or shifts in future shifts driven by changes in policy, environmental, or economic conditions that could constrain or render obsolete certain solutions. Using a filtering process, such as the one outlined in Appendix C, and carefully considering longer-term, slower moving changes, such as those described in Chapter 4, can help increase the likelihood that solutions chosen today will remain applicable and effective in an uncertain future.

Appendix A. Candidate and Selected Metrics

This appendix describes a scoring rubric we used to assess metrics against the criteria described in Chapter 2 and provides an overview of how the selected metrics (and some related metrics we did not choose) scored on these attributes.

Scoring Attributes

Metrics can be scored in several different ways. In this appendix, we use one of two three-point scales, depending on the attribute.

The *resource intensiveness* of a metric depends on the required manpower, equipment, infrastructure, and overall costs associated with metric measurements. We assessed this attribute using a low, medium, or high scale, in which *low* indicates the ability to readily collect applicable information for metrics with minimal to no additional costs, and high indicates that significant investments would be required to assess and track the metric being considered.

The second scoring method assesses the attributes of *validity*, *policy relevance*, *maturity*, and *operational usefulness* according to whether the metric fully meets, cannot meet, or partially meets the given attribute, as described in the following subsections.

Fully Meets Attribute

A metric is assessed as fully meeting an attribute when the metric adequately satisfies the definition of the attribute. In general, this will mean that the data required for the metric are currently being collected, that the metric is obviously useful across stakeholders, and that it is directly relevant to the aspect of the energy system being assessed.

Cannot Meet Attribute

Metrics that cannot meet the defined attribute will likely be readily apparent. It should be noted here that metrics are assessed as not satisfying the attribute only if meeting the definition is not possible under any circumstances. As discussed in the next subsection, however, if further investment might satisfy the attribute, the metric will be assessed as partially meeting the attribute.

Partially Meets Attribute

It will be obvious in a number of instances whether a metric can fully meet or does not meet the attribute at all. However, in a number of other instances the attributes may be partially fulfilled with a metric, and the underlying factors in these instances can be categorized.

For requirement metrics, partially meeting an attribute tends to point to instances of either incomplete information or the available information not being of the appropriate fidelity to fully meet the attribute. For instance, both the critical and nominal demand metrics partially meet the attribute related to setting targets because knowledge about future requirements is inherently incomplete. While a mission owner may be able to estimate power requirements from historical observations, these observations inherently incomplete knowledge in using these observations for future requirements. Additionally, the metric for nominal demand only partially meets the attribute relating to systematic Air Force collection. In this instance, the fidelity of current information is not at an appropriate level for decisionmaking. Currently, basewide power use can be approximated using monthly energy bills. While this can provide insights into seasonal variation and average power consumption, this does not capture the dynamics for shorter periods, making it very difficult to set requirements for peak demands.

Capability-based metrics tend to partially meet the attributes when detailed modeling and simulation or physical testing is required to understand the impact under future scenarios. In general, this is evident in the ability to collect metric information systematically. As an example of this case, consider the power supplied metric. Understanding how this capability will change in future scenarios will require either detailed modeling and simulation of the entire energy system architecture or extensive physical testing. Because it is not immediately apparent that doing either is feasible, the attribute cannot be fully satisfied. Additionally, the power quality metrics only partially meet the attribute for scalability in time and geography. These are special cases, though, and relate to the definitions of the metrics themselves. The proposed power quality metrics both tend to manifest locally (within a facility), so scaling with geography is not fully applicable. Also, total harmonic distortion (THD) is a continuous-state problem, and the voltage sags-and-swells metric is momentary (lasting up to 1 minute), so the metric is predefined with that time scale in mind.

Finally, for the proposed performance-only metrics, critical load not served and nominal load not served, the ability to meet the attributes relies fully on the parameters that make up the metric. Since there are instances in which the component parameters, demand and power supplied, only partially meet an attribute, this metric will also only partially meet that attribute for the reasons attributed to the component parameters. For this reason, we do not discuss the performance-only metrics in detail here.

Critical Demand

Critical demand is defined as the power required to sustain mission-essential functions (measured in kilowatts). Mission owners will likely measure the power requirement for individual missions, but the overall, basewide demand can be aggregated to any level useful for planning. Ultimately, the resolution of critical demand will depend on the power dynamics of the mission functions. Here, resolution is taken to encompass the scale of both the time and the geographic extent of the required measurements. For stable, predictable loads, longer-term energy use can be implemented to estimate the critical demand over extended periods. In these cases, monthly energy use may be used as a means of determining the average load during prescribed periods. However, many mission power requirements will likely vary over shorter periods and may include extreme peak demands. For these more-variable loads, the data required will require greater fidelity. It is expected overall that increasing the resolution of data gathered will aid in better planning.

We selected critical demand as a requirement measure for the level of power service because of the simplicity of this measure, which we expect will be explainable to all stakeholders, and because mission owners can set the targets directly. There may be some difficulty in setting targets for this measure if mission owners are not familiar with their current power requirements; however, this limitation in knowledge can be overcome with the collection of data at an appropriate resolution. While the Air Force may not be collecting data for this measure systematically now, investment in smart meters can provide this ability. Finally, this measure will be useful in planning and operations because it provides the power requirements directly. Table A.1 presents our evaluation for the critical demand metric.

Table A.1. Critical Demand Metric Evaluation

	Attributes	Critical Demand
	Type of metric	Requirement
	Validity	●
Policy relevance	Explainable to stakeholders	●
	Targets can be set directly	◒
Maturity	Systematically collected by Air Force	◒
	Systematically collected elsewhere (best practices)	●
Planning and operation	Useful in systems planning and real-time operations	●
	Scalable in time and geography	●
Resource intensiveness (low, medium, or high)		Medium

● Fully meets
◒ Meets with some difficulty
○ Cannot meet

Nominal Demand

The nominal demand is the power required to maintain normal base functions (measured in kilowatts), including mission-essential elements and other functions, such as housing. As with the critical demand, this measure can be aggregated to any level appropriate for planning but will likely be more applicable to the base civil engineer. Observations at existing bases suggest that the nominal demand will vary with respect to time of day and season. Long-term energy use, such as monthly energy billing, can provide some insight into the seasonal variation in power requirements, but daily variability will require finer resolution. Higher-resolution data can be collected using the same technologies proposed for collecting the critical demand. It is important to note, however, that the resolution at which the data are used for the energy system assessment (e.g., across a facility or base) is the scale at which measurements should be provided.

As with the critical demand, we selected nominal demand as a requirement measure because it is simple, explainable, easy to measure, and useful for planning and operations. Nominal demand is included because critical demand does not capture overall base functionality, and the importance will depend on the duration of outage scenarios. Table A.2 presents our evaluation for the nominal demand metric.

Table A.2. Nominal Demand Metric Evaluation

	Attributes	Nominal Demand
	Type of metric	Requirement
	Validity	●
Policy relevance	Explainable to stakeholders	●
	Targets can be set directly	◒
Maturity	Systematically collected by Air Force	◒
	Systematically collected elsewhere (best practices)	●
Planning and operation	Useful in systems planning and real-time operations	●
	Scalable in time and geography	●
Resource intensiveness (low, medium, or high)		Medium

● Fully meets
◒ Meets with some difficulty
○ Cannot meet

Power Supplied

Power supplied is the power transmitted to the system of study (measured in kilowatts). That system will likely coincide with the resolution of the aforementioned power requirements and may include a single facility, all mission-essential infrastructure, or the entire base. Determining the power supplied under different outage scenarios will require detailed knowledge of the base architecture and its functionality. Evaluation of the power supplied will thus likely require detailed modeling and simulation or some form of physical testing at the resolution required. When evaluating this metric, it is crucial to match the highest resolution of either critical load or nominal demand because direct comparisons between these metrics are made to determine performance gaps.

While power supplied is the proposed metric, it should be noted that other, similar metrics exist, such as derated power. *Derated power* is a measure of the expected available power but captures only the external power supply reliability and does not consider on-base backup generation. The main reasons for choosing power supplied over derated power are the ease of measurement, scalability, and usefulness in planning and operations decisions.

Power supplied is a capability measurement for the level of service. Because this is a capability metric, it is not evaluated with regard to setting targets. However, the power supplied has been determined to fully meet metric attributes related to policy relevance and planning and operation. This metric has some limitations in terms of systematic collection and ease of measurement because physical testing or detailed modeling and simulation would be required to explore potential future outage scenarios. Despite this limitation, further analysis of base energy architectures is expected to provide the information required to evaluate this metric. Table A.3 presents our evaluation for the power supplied metric, as well as derated power for comparison.

Load Not Served

Load not served is a measure of the performance gap between the power supplied and power demands. This means that load not served can be assessed for both the critical demand and nominal demand, depending on the desired energy system assessment. Because load not served is measured directly from the other proposed power metrics, its resolution will match the lowest resolution of the metrics included in its calculation. As with the previous level-of-service metrics presented, the appropriate resolution of this metric will be dependent on the power dynamics of the system of study.

While the load not served provides a time history of the power capability gap, a related metric, energy not supplied, provides similar information aggregated over a given performance period. *Energy not supplied* is the total energy gap between the required and supplied energy over a given period and can be calculated directly from load not served through integration over

Table A.3. Power Supplied Metric Evaluation

	Attributes	Power Supplied	Derated Power
	Type of metric	Capability	Capability
	Validity	●	◒
Policy relevance	Explainable to stakeholders	●	◒
	Targets can be set directly	N/A	N/A
Maturity	Systematically collected by Air Force	◒	○
	Systematically collected elsewhere (best practices)	◒	◒
Planning and operation	Useful in systems planning and real-time operations	●	○
	Scalable in time and geography	●	○
Resource intensiveness (low, medium, or high)		Medium	High

● Fully meets
◒ Meets with some difficulty
○ Cannot meet

a given period. We did not choose to use energy not served here because it provides aggregate information at a different resolution from the input measures and is therefore less useful in system planning and operations decisions.

The load not served is a performance measure for the level of service. Since it is not used to set requirements, it is not evaluated against the attribute for setting targets. As with the power supplied metric, load not served has been evaluated to fully meet the attributes pertaining to policy relevance and usefulness in planning and operation. The metric has maturity limitations that are due to difficulties in determining the power supplied under potential outage scenarios. Table A.4 presents our evaluation for the load not served metric.

Time to Restore Critical Functions and Nominal Operations

Restoration time is a measure of the duration from power loss until power demand is restored. As with power demand, restoration time can be measured for both the mission-essential functions and nominal operations at a base. *Time to restore critical functions* considers only the duration to restore mission-essential functions, which may be accomplished by moving the mission to another location. *Time to restore nominal operations*, on the other hand, includes the

Table A.4. Load Not Served Metric Evaluation

	Attributes	Load Not Served
	Type of metric	Performance
	Validity	●
Policy relevance	Explainable to stakeholders	●
	Targets can be set directly	N/A
Maturity	Systematically collected by Air Force	◒
	Systematically collected elsewhere (best practices)	◒
Planning and operation	Useful in systems planning and real-time operations	●
	Scalable in time and geography	●
Resource intensiveness (low, medium, or high)		Medium

● Fully meets
◒ Meets with some difficulty
○ Cannot meet

entire outage duration until normal functions are restored for the system of study. In general, the applicable resolution for these restoration times will depend on the requirements stakeholders set and may range from seconds to days. This information can be gathered from mission owners through questionnaires.

Related time and availability metrics exist, including the System Average Interruption Duration Index (SAIDI) and five-nines reliability. SAIDI is based on historical observations at the regional level and thus may not be applicable across the geographic scales considered here. Further, because SAIDI relies on historical observations, analysis involving low-probability, high-consequence outage scenarios is not possible directly with SAIDI. Five-Nines reliability is a measure of the expected availability of a system over a given period (generally a year). This availability measure requires detailed probabilistic knowledge of all system components and is difficult to calculate for low-probability outage scenarios. Further, the information on availability is aggregated over a longer period than the expected resolution for this problem, which makes this metric less applicable than direct time measurements.

The restoration time metrics proposed here are both requirement and capability metrics because they can be used directly in both applications. Additionally, the energy system assessment can evaluate the performance of the system with respect to time and availability by calculating the gap between the requirement and capability. In general, restoration time has been determined to fully meet attributes pertaining to policy relevance and usefulness in planning and operations. It has been determined that there is some difficulty in meeting maturity attributes, especially as a capability metric, because of the need for detailed modeling or testing of the system architecture. These limitations are similar to those discussed for power supplied. Table A.5 presents our evaluation for the time to restore critical functions and time to restore nominal operations metrics.

Cost of Outage

The cost of an outage is an incremental measure of the increased costs to operate under a given outage scenario. As previously discussed, it will likely include costs associated with fuel use, manpower, and equipment maintenance. Further, when significant changes to the energy system architecture are required, these costs may also be assessed based on all associated infrastructure investments. The ultimate goal of including this cost metric is to provide monetary valuations for cost-benefit or cost-effectiveness assessments. Base civil engineers will likely evaluate these costs, with input from mission owners.

Table A.5. Restoration Time Metric Evaluation

	Attributes	Time to Restore Critical Functions	Time to Restore Nominal Operations	System Average Interruption Duration	Power Availability
	Type of metric	Requirement and capability	Requirement and capability	Capability	Requirement and capability
	Validity	● ●	● ●	○	○ ○
Policy relevance	Explainable to stakeholders	● ●	● ●	◒	◒ ◒
	Targets can be set directly	● N/A	● N/A	N/A	○ N/A
Maturity	Systematically collected by Air Force	◒ ○	◒ ○	○	○ ○
	Systematically collected elsewhere (best practices)	● ◒	● ◒	◒	○ ○
Planning and operation	Useful in systems planning and real-time operations	● ●	● ●	○	○ ○
	Scalable in time and geography	● ●	● ●	○	○ ○
Resource intensiveness (low, medium, or high)		Low—High	Low—High	N/A	N/A

● Fully meets
◒ Meets with some difficulty
○ Cannot meet

While this metric was the only measure evaluated in the cost category, it has been determined to sufficiently meet all attributes associated with the energy system evaluation. This is expected; cost is often included in this type of analysis. Table A.6 presents our evaluation for the cost of outage metric.

Total Harmonic Distortion

The first of the two power quality metrics we propose is THD, which measures the contribution of all harmonic frequency currents to the fundamental frequency (60 Hz). Total harmonic distortion is a continuous state problem arising from the operation of equipment with nonlinear current draw. Harmonics within an electrical system tend to vary throughout the distribution system, and analysis for total harmonic distortion is therefore required at fine geographic resolutions. This typically occurs at the electrical panel or at the busses.

THD provides a measure for both the requirements and capabilities of the energy system. In general, it has been assessed that THD fully meets the metric attributes related to explicability and usefulness in systems planning and operation. However, using THD as both a requirement and capability has some limitations in most other metric attributes. For requirements, this is due to the difficulties associated with understanding specific equipment operating limits for total

Table A.6. Cost Metric Evaluation

	Attributes	Cost
	Type of metric	Performance
	Validity	●
Policy relevance	Explainable to stakeholders	●
	Targets can be set directly	●
Maturity	Systematically collected by Air Force	●
	Systematically collected elsewhere (best practices)	●
Planning and operation	Useful in systems planning and real-time operations	●
	Scalable in time and geography	◒
Resource intensiveness (low, medium, or high)		Medium

● Fully meets
◒ Meets with some difficulty
○ Cannot meet

harmonic distortion. Fully understanding these limits would require physical testing or detailed manufacturer information. On the capability side, the limitations in THD are due to the lack of currently collected data at the resolution required. This capability-side limitation can be overcome through investments in power quality metering equipment. Table A.7 presents our evaluation for the THD metric.

Voltage Sags and Swells

Measurement of voltage sags and swells is another power quality measure that provides valuable information over short time durations. *Voltage sags* are momentary reductions in the root-mean-squared (RMS) voltage of 10 to 90 percent of the nominal voltage over a duration of one-half cycle to one minute. The momentary increase in RMS voltage is known as a *voltage swell*. Generally, these sags and swells are localized problems and are due to short circuits, overload, or the starting of electric motors. Because these voltage changes are localized and brief, measurements are required at a fine geographic and time resolution.

Table A.7. Total Harmonic Distortion Metric Evaluation

	Attributes	THD	
	Type of metric	Requirement and capability	
	Validity	●	●
Policy relevance	Explainable to stakeholders	●	●
	Targets can be set directly	◒	N/A
Maturity	Systematically collected by Air Force	◒	◒
	Systematically collected elsewhere (best practices)	◒	◒
Planning and operation	Useful in systems planning and real-time operations	●	●
	Scalable in time and geography	◒	◒
Resource intensiveness (low, medium, or high)		Medium	

● Fully meets
◒ Meets with some difficulty
○ Cannot meet

A related metric, in terms of effect to end use equipment, is *flicker* which refers to the perceived change in brightness of a lamp due to the rapid voltage fluctuations. It has been noted in the literature that analytic determination of flicker may not always be possible and that it is extremely difficult to base requirements for sensitive equipment on this measure. For these reasons, measures related to voltage sags and swells are preferred.

Voltage sags and swells can be used as both requirements and capabilities in assessing the energy system. As with THD, voltage sags and swells should perform well in metric attributes related to explainability and usefulness in systems planning and operation. However, setting requirements based on this metric will require detailed knowledge of how sensitive end-user equipment is to short-duration variations in voltage. This limits the metric in setting targets and ease of measurement for requirements. Additionally, relevant voltage information will need to be collected systematically at fine resolutions, which may not be done currently. These measurement limitations may be overcome through investments in the same type of power quality measurement equipment required for THD. Table A.8 presents our evaluation for the voltage sags and swells metric.

Table A.8. Voltage Sags and Swells Metric Evaluation

	Attributes	Voltage Sags and Swells	Flicker
	Type of metric	Requirement and capability	Requirement and capability
	Validity	● ●	○ ○
Policy relevance	Explainable to stakeholders	● ●	◒ ◒
	Targets can be set directly	◒ N/A	○ N/A
Maturity	Systematically collected by Air Force	◒ ◒	○ ○
	Systematically collected elsewhere (best practices)	◒ ◒	○ ○
Planning and operation	Useful in systems planning and real-time operations	● ●	○ ○
	Scalable in time and geography	◒ ◒	○ ○
Resource intensiveness (low, medium, or high)		Medium	Medium/ high

● Fully meets
◒ Meets with some difficulty
○ Cannot meet

Appendix B. Additional Framework Implementation Guidance

This appendix demonstrates how users would walk through the problem identification portions of the framework outlined in Chapter 5. We do not go into detail here on the process of selecting and implementing the appropriate set of solutions to address those problems because that process is bound to be case specific. Appendix C lists candidate solutions that could be applied to different types of gaps and presents an approach for selecting appropriate solutions.

In the sections that follow, we provide an example application of the process in Figure 5.1, broken down into five steps. For each step, we describe who would need to be involved, what information they would need to acquire and process, and how they would likely accomplish the step. We developed this walkthrough by drawing on inputs and feedback received from mission owners and civil engineers at several CONUS installations. Additionally, we conducted an informal in-person exercise at one installation to get the practitioner's perspective on the usefulness of the RAND framework and what it would take to implement it.

Table B.1 summarizes these five steps, which involve gathering data and gaining an understanding of the building blocks of system architecture, requirements, and capabilities.

These five steps are essential to establishing the foundation of the framework. Mission owners and engineers need a common understanding of and perspective on the elements of system architecture, baseline capabilities that can meet requirements under normal operations, and the consequences of scenarios that degrade capabilities and lead to failure to meet

Table B.1. Steps Involved in the First Part of the Framework (Figure 5.1)

Framework Component	Who Needs to Take the Lead	What Needs to Be Done	How to Accomplish the Task
System architecture	Base civil engineer and mission owners	Identify and communicate elements of system architecture	Meet face to face, execute checklist, visit sites
Scenario conditions	Base civil engineer	Assess capabilities—How might a scenario impact supply of and demand for power?	Discuss, model, simulate
Requirements	Base civil engineer and mission owners	Identify and communicate requirements	Meet face to face, execute checklist, visit sites
Base capabilities	Base civil engineer	Identify potential gaps between capabilities and requirements (Is there a problem?)	Discuss, model, simulate
Outcomes	Base civil engineer and mission owners	Assess outcomes (money, loss of mission functionality) of degraded capabilities	Discuss, model, simulate

requirements. In general, mission owners and base civil engineers will need to converge on a mutual understanding of base architecture, capabilities, and requirements.

Identify Elements of the Base Electrical System Architecture

The first step in the framework is for all involved to understand the system architecture, essential for assessing the level of energy assurance at a base. Table 5.1 provided an overview of key system architecture components. The base civil engineer with help from the base civil engineering team (including energy planners) should begin by documenting attributes of the primary sources of power on the base:

- Who is the utility provider, and how much (if any) power is produced on the base?
- Are there any contracts or priority agreements with the utility? What do they cover? For example, if there is a power outage, does the base have priority in having power restored?
- Has the utility provided information about the likelihood of power outages, or does the utility send advance warning of rolling outages on high demand days? How are mission owners alerted to such events?
- Are recent or upcoming changes in the local power grid likely to affect the reliability of the power grid?
- Where are single points of failure associated with access to the power grid? (For example, does a single substation or a single, aging high-voltage line serve a large portion of the base?) How difficult (in terms of time and money) would it be to fix one of these critical primary power assets were they to fail?

Next, the base civil engineer and mission owners should identify and document attributes of backup power sources and power conditioning equipment on the base:

- How many fixed and mobile generators are on the base? What is their fuel consumption?
- How frequently are they tested, and how frequently do they have problems?

In particular, mission owners should identify any elements of system architecture that they own, as distinct from other elements of basewide architecture:

- Does the mission owner have additional generators or UPS systems attached to their facility?
- Who is responsible for maintaining those systems, and how frequently do they experience mechanical issues?

Both base civil engineer and mission owners should have an understanding of the reliability of both primary and backup power sources.

Diesel fuel is critical to most backup generators. The base civil engineer and mission owners should discuss how much fuel is stored at various sites, who is responsible for replenishing it, and how frequently. The base civil engineer and mission owners should confirm their understanding of whether jet fuel is an appropriate alternative fuel source for their generators. Some downrange generators are designed to use jet fuel as an input. However, using jet fuel instead of diesel can rapidly damage many commercial generators. Discussions should also include questions relating to the repair of backup systems, such as the following:

- Are parts and equipment available on base?
- Are only contractors permitted to service certain systems?
- What factors influence how quickly a backup system can be repaired?

Information characterizing system architecture elements rarely resides in one office or with one individual, if the information exists at all. Table 4.1 indicated which entities are likely to own data related to the various aspects of system architecture.

Identify Requirements

Just as mission owners need to understand what resources are available, the base civil engineer should understand the needs of mission owners that constitute the demand for the energy the base architecture provides. This is particularly important for helping the base civil engineer correctly prioritize resources in the event of an outage. Mission owners should be able to identify and differentiate between nominal and mission-essential power needs. Mission owners should also identify whether and when limited base capabilities might alter the power requirements for critical missions. For example, many critical missions have standard procedures for moving to an alternative location if an outage lasts beyond a preset duration. If the base is evacuated, the entire mission may be moved, or the requirements might be reduced to enable support of a minimal operation.

Table 5.3 identified broad categories of capabilities and requirements that the base civil engineer and mission owners should discuss. Discussion might begin by identifying which facilities are associated with each mission and what the power requirements are for that building under normal conditions. Does the mission use more energy on hot days? Many missions have sensitive equipment that must remain cool. Does the mission use more energy during the day or at night? Some missions may have more staff on site during the day consuming energy, while others may need more energy at night for lighting. Mission owners should also discuss how stable or variable their demand for energy is. Some missions keep a continuous level of demand 24 hours a day, while others may demand large amounts of energy in short bursts. Mission owners should also be aware of and communicate to the base civil engineer how capable their mission is of reducing energy demand when given advance notice, such as for days where rolling blackouts are more likely. Some missions may be able to move high energy consumption to alternative days or times, while others are required to operate constantly, without interruption. Mission owners should be clear about what duration of outage is problematic for completing their mission requirements. If the duration of outage that would be problematic varies, mission owners should explain what factors cause the variance. For example, an hour-long outage in the afternoon may be severely problematic, while an hour-long outage at midnight may be less problematic.

Mission owners should think through interdependencies between electricity and other mission-essential infrastructures in assessing and communicating power requirements. For example, water infrastructure (e.g., pumps and flood gates) is often tied directly to the electric grid. A power outage could also mean water or communication systems are unavailable.

Assess Capabilities

Once the system architecture and requirements have been identified and communicated, the next step is to identify the particular combination of capabilities and requirements that result from a particular scenario. For this step, the base civil engineer and staff should consult Table 5.2, which provided examples of five different event-driven scenarios that could affect a base's capabilities. Focusing on one scenario in particular, and potentially modifying the narrative to better reflect their base's unique risks, the base civil engineers and their staffs should consider how that scenario would affect the base's energy capabilities. Table 5.4 highlighted some examples of linkages between system architecture, scenarios, and capabilities. In a team setting, the staff could brainstorm the potential impacts of the scenario. If possible, staff with long-term institutional memory should be included; their knowledge of how historical scenarios affected base architecture would be extremely helpful. Recording and maintaining this institutional knowledge is valuable for base resilience. For example, the base civil engineering staff may keep records on the details of prior outage events.

The mechanisms through which a scenario might determine capabilities on a base can be quite complex. Capabilities are defined by metrics, such as power supplied. If the task at hand is to determine power supplied in Scenario E3 from Table 5.2 (an ice storm or a hurricane like Sandy), a starting point would be to list all sources of backup power and identify how much power each can supply and to which facilities and missions. Next, the potential impacts of physical effects, cybereffects, and power quality events should be considered. For Scenario E3, the main concern is physical effects. The potential impacts of a flood, ice storm, or other physical event on each backup system should be considered. For example, are any generators in low-lying areas and unprotected from rising waters? Flooding could make these systems unable to function, so they would no longer contribute to the power supplied in this event. The impact of the scenario on power supplied is not limited to direct damage to generators. Flooding could close roads, making fuel deliveries difficult or impossible. A lack of fuel could severely restrict the ability to provide power from all diesel generators for the entire duration of the outage. Base personnel and off-site contractors may be unable or unwilling to come to the base because they are busy dealing with problems at their own homes caused by flooding. The lack of personnel could make fixing broken systems a slower process or completely impossible.

Each of these issues reduces power supplied from the starting point of all backup systems providing power for the entire duration of the scenario to subsets of the systems providing generation for various lengths of the outage. For example, suppose one-third of all backup systems are directly damaged from flooding, and the lack of personnel means these systems cannot be repaired promptly. Further, suppose flooding has severely damaged a key bridge, making fuel deliveries impossible. This base now only has the capability to supply power using two-thirds of its normal backup systems and can only run the systems for a few days. Modeling and simulation or physical testing might be required for a precise understanding of whether and to what extent a particular scenario would degrade capabilities on a base, including specificity about which missions would be affected.

Our test run of the framework at one base indicates that these discussions are best supported by having both Table 5.1 (inputs to system architecture) and Table 5.2 (event-driven scenarios) available for consultation.

Identify Potential Gaps Between Capabilities and Requirements

For each scenario, the base civil engineer should consider how the ability to provide power to the base matches the requirements that will be demanded from mission owners in this scenario. If there is any uncertainty about what requirements would be in that scenario for a particular mission, that mission owner should be contacted. The ability to have open communication over such questions is valuable and best resolved in advance. Attempting to guess mission needs for a particular scenario in real time could delay or degrade response times.

Assessing gaps between requirements and capabilities is not easy. Even if the capabilities in the particular scenario are carefully assessed as part of the previous step, requirements from the missions might change depending on the scenario. Continuing the example of Scenario E3, nonessential base personnel may be evacuated in advance of the natural disaster. To calculate requirements for power supplied, the base civil engineer would need to know which buildings are being evacuated and whether there will be any residual energy demand from those buildings. For example, evacuations in winter conditions might still require heating systems to operate at a minimal level to prevent pipes from freezing. Learning what power would be demanded in this situation requires speaking with the facilities manager at each building site.

Further, many missions have contingency plans for shifting operations to alternative locations if outages last more than a certain length of time. Again, shifting operations may not reduce demand for power to zero; there may be residual demand for power from systems that cannot be moved or systems that still require minimal staff support. This residual demand for power supplied may still be mission essential because it may prevent damage to expensive equipment. Again, understanding the power needs in this setting may require speaking with the facility manager, and understanding the critical nature of any residual demand may require speaking with both the facility manager and mission owner.

Once the capabilities and requirements, and how they change over time, are understood for the scenario, the differences between them can be calculated. It may be the case that capabilities exceed requirements for the first several days of a scenario but that requirements exceed capabilities as time wears on. If requirements exceed capabilities at any point in time, this creates a risk that should be documented. Further complications are also possible. For example, perhaps capabilities are impaired in such a way that power supplied exceeds requirements on part of the base, but requirements exceed capabilities on another portion of the base. If the base architecture is such that sufficient excess capability cannot be moved from one portion of the base to the other, this also represents a risk that should be documented.

Once the system architecture and requirements have been identified and communicated, the next step is to identify the particular combination of capabilities and requirements that results in a particular scenario. For this step, base civil engineers and their staffs should consult Table 5.2, which defined the five different event-driven scenarios that could affect a base's capabilities. Focusing on one scenario in particular, and potentially modifying the narrative to better reflect a base's unique risks, base civil engineers and their staffs should consider the impacts of that scenario on the base's energy capabilities. Table 5.4 highlighted some examples of linkages between system architecture, scenarios, and capabilities. As with assessing capabilities, staff should brainstorm potential scenario impacts, drawing in particular on personnel with long-term institutional knowledge of how historical scenarios affected base architecture. Our test run of the framework at one base indicates that these discussions are best supported by having both Table 5.1 (inputs to system architecture) and Table 5.2 (event-driven scenarios) available for consultation.

Assess Outcomes

Once potential capability gaps for each scenario are identified in the previous step, each gap should be evaluated for (1) its specific impact—or outcome—on a mission, in the context of duration, and (2) the total marginal costs associated with operating through a scenario. Careful evaluation of outcomes can help narrow the response option space and guide more-effective investments to close energy assurance gaps. Appendix C proposes a process for linking gaps and outcomes to response options.

Continuing the example of Scenario E3, suppose downed power lines and debris interfere with base capabilities to meet the mission restoration time requirement and that the mission owner chooses to move the mission to another installation. One direct outcome in this case is the unexpected need to move the mission. But to understand the implications of this outcome fully, the base civil engineer and mission owner should consider both direct and indirect impacts, and their associated costs. One direct impact to consider is the new power requirement at the base taking over the mission. For a short time, that base may be able to meet the new power requirement with no problem. However, at some point, this new (emergency) power requirement could tax base capabilities to the point that the base civil engineer and mission owner must deal with competing power demands at that base, with the potential for additional direct or indirect impacts on the mission at one or both bases. Mission owners and base civil engineers should consider all these impacts in determining the total marginal costs of operating through a scenario.

Some key questions to help tabulate the marginal costs associated with operating through a scenario include the following:

- What resilience actions did the base civil engineer take?
- What resilience actions did the mission owner take?
- Was any sensitive mission-critical equipment damaged?
- If the mission was moved to another base, how was this performed? Was there a physical handoff? What actions did the receiving base take to accommodate the mission?
- Did any of the resilience actions require additional manpower? Overtime? Equipment or facilities?
- Were any additional emergency power supplies procured?
- How much fuel did backup power generators consume?
- What mission-essential functions could not be performed during the outage? For how long? Did this result in mission failure?
- What other base functions were affected by the outage (e.g., water supply, communications, housing), either directly or indirectly, because of competing power demands? For how long?

The ultimate goal of considering this cost metric is to provide monetary valuation for benefit-cost or cost-effectiveness assessments. If outage costs are less than solution costs across scenarios, maintaining the status quo of base architecture and capabilities is likely to be the preferred action.

While we have identified key players for each framework step discussed in this appendix, it is important to have a designated person or a group of people who would *own* the process of working through the steps in the framework. Also, while data or information collection does not necessarily need to be centralized, access to data and information needs to be easy. Using a common portal that all relevant entities on a base can use to feed and access energy data can help make sure that data and information are available to those who need it, when they need it, and in a standard format.

Appendix C. Identifying, Analyzing, and Implementing Response Options

Broadly speaking, this report has presented a framework for assessing and planning to ensure that Air Force bases have access to sufficient electric power to perform their missions. Part of that framework is identifying and addressing gaps between energy requirements and capabilities. This appendix uses the set of response options we identified in this research and focuses on the process of identifying potential responses for specific gaps and selecting the best ones for the circumstances.

After introducing gaps, we present and describe the full set of response options we developed. Next, we describe our process for mapping response options to gaps to define a reference set of candidate options. To determine which options from the reference set may be helpful in a given situation, we next offer a series of filters and demonstrate how to apply them. Then we discuss a set of categories of action—*levers*—and the responses that fall into these categories. Using these levers to classify workable options helps illuminate potential pathways for resourcing response options. Finally, we describe how a list of candidate options can be sorted into tiers to determine which ones are most feasible given available resources and regulatory constraints.

Gaps

To determine an appropriate solution to a problem, one must first identify the problem, represented in this work as a *capability gap*. The framework presented in Chapter 5 can be used to identify capability gaps a given installation may experience that affect its ability to withstand, respond to, or recover from a power disruption in a given scenario. By way of illustrating how metrics can be used to identify gaps, that chapter focused on three broad categories of metrics: time to restore (critical functions and nominal operations), load not served (both critical and nominal), and issues with power quality (voltage sags and swells, THD).

Because we dealt with illustrative scenario-based problems here, we used these broad gaps, rather than identifying specific problems, to help decisionmakers classify problems and solution sets in scenarios when all the data may not be available. Having solutions prepared to address a broad type of energy gap permits more-robust planning than choosing solutions based on narrow problems.

Each gap the framework identifies may have more than one root cause, depending on the base system architecture and the scenario conditions. The nature of the specific problem will influence which response or set of responses is most appropriate for addressing the problem. The framework presented in this work identifies only the candidate set of response options from which this selection can be made; it does not identify the best response. Additionally, our analysis considered each option independently; we did not address combinations of responses, although a combination of responses may be technically viable in some cases. In subsequent sections, we will explain our methodology for mapping options to gaps and the criteria used to determine the candidate set of responses.

Response Options

We attempted to capture options that improve performance under outage scenarios and/or that minimize risk to critical mission systems. To arrive at the full set of response options, presented in Table C.1, we drew on the team's technical expertise, supplementing that with information from the literature on energy technologies and power-distribution infrastructure.

Our final set of response options includes both *materiel* solutions, such as investing in on-site prime power generation technologies and infrastructure investments that may help improve reliability, and *nonmateriel* solutions, such as reassigning personnel or making changes to emergency procedures that will result in more rapid recovery during an outage.

Table C.1. Response Options

Option	Description
Add Air Force personnel	Includes calling in off-duty Air Force personnel, reassigning personnel from other organizations on base during an outage, and adding manning authorizations. May be temporary or permanent changes to assignment or positions.
Automate power distribution for critical load	Technology that automates the ability to control power distribution to critical loads. Involves investment in a switching and control mechanism.
Add automatic transfer switch for commercial to backup generation	Technology that automates switching from a commercial power bus to a generator bus in the event of a power disruption on the commercial line.
Add batteries	Addition of battery storage for backup power during an outage.
Add biomass power-generation plant	Use of biomass as a fuel source in direct combustion; biomass is burned, and the heat generated is used to create steam, which turns a turbine to generate electricity. Typically prime power generation.
Add combined heat and power (CHP) generation plant	Use of a heat engine or power station to generate electricity. Recovered waste heat is used in other processes within the power plant, such as a steam turbine or another industrial process. Typically prime power generation.
Perform systematic monitoring	Data collection and analysis on a regular basis that may assist in identifying system vulnerabilities or problems before they occur. Includes data collection associated with framework metrics to help in determining requirements. Most effectively performed with electronic systems, such as an energy monitoring and control system (EMCS), smart meters, or other supervisory control systems.
Contract out emergency services	Use of contractual mechanisms to augment existing resources for emergency services. Can include personnel not on a traditional contract, access to spare parts, or access to non–Air Force resources on short notice.

Table C.1—Continued

Option	Description
Add cybersecurity controls	Implement cybersecurity controls for industrial control systems, specifically critical power-generation equipment, to minimize the risk of mission failure in the event of cyberattack. These controls may include implementing access controls, isolating systems, monitoring network activity, sniffing for wireless access ports, and scanning removable drives for malware before connecting to the system.
Implement demand response, peak demand shaving, or load shedding	Reducing power demands during peak times or during an outage to lessen loads on electrical lines and/or ensure that critical loads can be served with a reduced power supply. Includes shifting deferrable loads to another time (i.e., off-peak times).
Add diesel power generation	Add electricity generation using a diesel generator. Typically backup or emergency power generation.
Add electric vehicle to grid	Add electric vehicles specially designed to plug directly into the base power-distribution system that can act as an emergency battery supply.
Add EMCS	Add a system of sensors, transmitters, data acquisition, and data processing that can be performed at the user (i.e., building) level. May also include data and control systems that are full installation control schemes.
Add energy storage (other than batteries)	Add an energy storage system, other than batteries, to be used in combination with power-generation equipment. Examples include thermal energy storage, flywheels, compressed air energy storage, and chemical storage.
Add fuel cell	Add technology that converts hydrogen-rich fuels to electric power and waste heat. Some fuel cells are able to reuse waste heat internally in the process of converting the fuels; some can be paired with other technology that can capture and use the heat in another process. Fuel cells can serve as prime power generation for base loads or can serve smaller isolated loads, depending on the size of the fuel cell.
Increase liquid fuel storage capacity	Increase the storage capacity for liquid fuels needed for operation of backup or emergency power-supply equipment. This may include permanent storage tanks or temporary storage, such as fuel bladders.
Add geothermal power generation	Add technology that captures energy from the earth (typically by capturing steam) and converts it to electric power using a turbine. Typically prime power generation. Includes dry steam, flash steam, and binary cycle power-plant designs.
Add islanded-mode enabled microgrid	Add a collection of technologies that form a small-scale, local power grid that can operate in parallel with or independently of the utility grid. When properly configured, microgrids may provide stand-alone power in the event of utility power disruption. Microgrids are often designed with both power generation and storage technologies. Microgrids can be any combination of technologies.
Add liquefied natural gas (LNG) power generation plant	Add technology that uses LNG to generate electricity through direct combustion using a turbine. Typically prime power generation.
Enable load shifting	The ability to shift an electrical load from one electrical line to another. May be performed with automated technology or manually.
Upgrade or repair substation	Upgrade or repair substations that feed power to the base.
Add methane power generation plant	Add technology that uses methane as a fuel source to generate electricity. Includes waste-to-energy generation, biodigestors, and landfill methane capture. Typically prime power generation.

Table C.1—Continued

Option	Description
Enable moving the mission	Backup or contingency measures to prevent or minimize disruption to critical mission tasks in the event of a power disruption. May involve handing off mission tasks to personnel at alternative locations or moving the mission to alternative locations until normal operations may resume at the primary location.
Move electrical lines underground	Process of moving overhead power lines underground to reduce risk of damage from natural or human causes.
Move critical power equipment from at-risk locations	Move critical power-generation equipment (primary or backup) from at-risk locations (e.g., remove generators from basements to prevent damage in the event of a flood).
Add structural hardening	Investment in physical hardening of structures that support or house critical power-generation equipment to prevent physical damage from natural disasters or human tampering.
Add physical security	Provide physical security (i.e., add manpower) to critical power-generation equipment to deter human tampering that could damage equipment or cause an outage.
Add power distribution unit (PDU)	Add technology designed to efficiently distribute electric power to multiple devices to improve reliability. Typically used in conjunction with uninterruptible power supplies for network equipment, in data centers, or for other sensitive equipment requiring high up-time.
Implement preventative maintenance schedules	Develop and implement preventative maintenance schedules for all prime power generation, backup power generation, and critical supporting infrastructure to ensure proper operation at the time of need. May also include general base maintenance, such as maintaining trees to prevent limbs interfering with or damaging overhead power lines.
Make procedural changes	Change the sequence of operations, for either mechanical equipment or procedural operations, to enhance recovery in the event of a power outage.
Improve quality of fuel-supply contract	Improve the quality of the fuel-supply contract to minimize risk of interruption of fuel deliveries (e.g., ensure the base is the primary recipient of any limited supplies, secure redundant sources of supply).
Add small nuclear power plant	Add technology that uses a nuclear reactor to generate heat that is used to create steam that powers turbines to generate electricity. Typically prime power generation.
Add solar photovoltaic power generation	Add technology that converts solar energy to electricity directly with photovoltaic technology. Not typically considered prime power generation because weather can cause output to vary.
Add UPS	Add technology that conditions power flow to sensitive electronic equipment and provides continuous power for a short period when the primary power source is disrupted. Typically used in conjunction with a PDU for network equipment, in data centers, or for other sensitive equipment requiring high up-time.
Train existing personnel	Train existing Air Force or contractor personnel to improve their ability to react and respond to outages.
Upgrade or replace utility lines	Upgrade and replace utility lines to improve the reliability of the power supply and quality.

Table C.1—Continued

Option	Description
Add wind power generation	Add technology that uses wind to mechanically generate electricity with a turbine. Not typically considered prime power generation because weather can cause output to vary.
Add redundant electrical lines	Add more electrical lines that can serve power to critical and nominal demands. Includes bringing additional utility lines onto the base for commercial power generation or adding lines to base infrastructure and/or individual facilities.

Mapping Response Options to Gaps

To begin the process of selecting responses for a given gap, we first identify potential root causes for each gap without regard for any particular base system architecture or scenario conditions. We begin with gaps instead of response options to simplify the mapping process because some response options may be suitable solutions for more than one gap. We then examine the full list of response options and select and justify possible responses using our own technical expertise to evaluate the purpose or benefit of each option. This produces a reference set of candidate responses for each gap. As an example, Table C.2 shows the results of this mapping for the gap *unacceptable restoration time*. The full set of response option–to-gap mappings is available in supplemental material that can be requested from the authors.

Identifying Candidate Solutions: Using Filters

Filtering through the reference set of candidate responses is a way of identifying which response or responses are most appropriate for the identified energy resilience or reliability challenges an installation faces by examining base system resources and scenario conditions to evaluate whether an option is feasible. In this section, we discuss and illustrate two filters that can further assess the suitability of candidate responses. These filters are based on the governing factors applicable to the situation.

Filters for Response Options

We attempted to capture the critical elements decisionmakers should consider when evaluating responses for investment or execution during an outage, presented here as *filters*. The *Baseline Resource Requirements* filter is based on governing factors that may help decisionmakers decide whether baseline resources and infrastructure will support a given option. These factors are presented in Table C.3. The *Execution During an Outage* filter is based on governing factors that relate to the scenario conditions of a given outage. These factors are presented in Table C.4. Before investing in solutions, decisionmakers should also consider a third filter: possible changes or shifts in future conditions that could constrain proposed solutions

Table C.2. Response Option–to–Gap Justification: Unacceptable Restoration Time

Potential Root Causes	Response Options	Justification
Personnel needed to turn on generators manually	Add Air Force personnel or contract out emergency services	Fills manpower need, increases performance and capabilities, and contributes to resilience
Unable to meet critical demand in appropriate amount of time	Enable moving the mission	Could reduce demand on the installation during the outage; contributes to resilience
Have enough people but they are not properly trained to run backup power supply equipment	Train existing personnel	Improves performance and recovery time, contributes to resilience
Unable to restore power in a timely manner because no cybersecurity countermeasure program is in place	Add cybersecurity controls	Could help respond to and recover from an attack; improves performance and capabilities and contributes to resilience
Fuel delivery is unreliable	Improve quality of fuel supply contract	May assist in bringing additional generation equipment online faster because the supply is more reliable; contributes to resilience and improves capability
Emergency response actions are not implemented in the most effective way, causing delays	Make procedural changes or add EMCS	For example, an automated control sequence could help personnel respond more quickly to an outage; contributes to resilience and improves performance
Not enough fuel to bring additional capacity online	Increase liquid fuel storage capacity	Having more capacity may help bring other backup generation online quicker because it may be sufficient to meet critical demands and nominal demands; contributes to resilience and improves performance and capabilities

or render them obsolete. Such shifts may occur slowly over time and are driven by changes in policy, environment, or economic conditions. This filter is discussed in more detail in Chapter 5.

Using the governing factors presented in Tables C.3 and C.4, we developed two sets of questions that correspond to the filters. Base civil engineers and planners can use these questions during the "Consider Response Options" and "Unresolved Capability Gap" stages of the framework to further assess the suitability of candidate responses. These questions were informed by the literature and both the Air Force's and the team's subject-matter expertise. As mentioned previously, our analysis considers each option independently; we did not address combinations of responses, although a combination of responses may be technically viable in some cases. Additionally, our analysis does not account for full life-cycle costs or mission requirements and assumes that the Air Force is investing Air Force dollars in the response, so we do not include questions about some contracting mechanisms, such as enhanced use leases and public-private partnerships.

Table C.3. Filter for Baseline Resource Requirements

Governing Factor	Description
Capability of off-base personnel	Adequate personnel are trained, available, and capable of performing an emergency function on base and must get on base. This includes Air Force and contractor personnel who have received prior approval to access and work on the base or in secure areas but does not consider physical restrictions to base access, such as a flooded road.
Capability of on-base personnel	Adequate personnel are trained, available, and capable of performing an emergency function and must be on base. Includes Air Force and on-base contractor personnel.
Control of O&M and equipment upgrades	Air Force control of equipment or personnel is required to execute the response. *Control* includes situations in which the Air Force can contractually obligate contractors to perform actions or comply with requirements.
Infrastructure investment	Investment in infrastructure is required before or during execution of a response.
Local, state, or federal regulatory approval	Local, state, or federal regulatory approval is required to execute the response (e.g., interconnection agreement with utility, permits, or an environmental impact assessment).
Viable site conditions	Specific site conditions are required to execute response option (e.g., solar requires sufficient sun; an LNG plant requires the ability to deliver LNG).
Viable site space	Physical space is required to execute the response. This does not consider site conditions, such as sufficient sun exposure for solar power generation.

Table C.4. Filter for Execution During an Outage

Governing Factor	Description
Access to base, facility, or equipment is available during outage	Physical access to the base, facility, or equipment is required for execution of the response. Physical access is available during the outage. This assumes that Air Force or contractor personnel have received prior approval to access and work on base or in secure areas.
Sufficient battery supply for duration of outage	The technology requires connection to a power source to be recharged and ready for redeployment when supplies are depleted. The stored battery supply is sufficient to function at required capacity for the duration of the outage. If the battery supply is not sufficient for the duration of the outage, there must be a connection to a power source to recharge.
Sufficient fuel supply for duration of outage	The response requires a fuel source to be functional. The fuel supply (existing storage or resupply) is sufficient for equipment to function at the required capacity for the duration of the outage.
IT/cyber systems are functional	The response uses IT/cyber systems for normal operation. These systems are functional during the outage. If they are not, equipment must be able to operate in manual mode.
Maintenance does not disrupt power supply	Equipment and critical supporting infrastructure are functioning for the duration of the outage. Maintenance to equipment or critical supporting infrastructure that would take the equipment off line or interrupt the power supply to assigned demands is not required during the outage. If maintenance is required that would bring the equipment off line or disrupt the power supply, another backup power supply must be provided.
Switching mechanism or inverter technologies are functional during outage	The response relies on a switching mechanism or inverter technologies for operation (e.g., manual and automatic transfer switches). The switching mechanism or inverter technologies are functional during the outage.

To illustrate the application of these filters, we will walk through an example using an islanded-mode enabled microgrid as the response option under consideration. First, to determine whether a response option is appropriate for a specific base or outage scenario, a planner should consider and identify all applicable governing factors from Tables C.3 and C.4. We present the applicable factors for an islanded-mode enabled microgrid in Tables C.5 and C.6. An *X* indicates

Table C.5. Islanded-Mode Enabled Microgrid Governing Factors for Investment Decisions

Governing Factor	Response Option: Add Islanded-Mode Enabled Microgrid
Capability of off-base personnel	X
Capability of on-base personnel	Situational: Does Air Force need additional personnel support?
Control of O&M and equipment upgrades	
Infrastructure investment	X
Local, state, or federal regulatory approval	X
Viable site conditions	Situational: Depends on technologies
Viable site space	Situational: Depends on technologies

Table C.6. Islanded-Mode Enabled Microgrid Governing Factors for Execution Decisions

Governing Factor	Response Option: Add Islanded-Mode Enabled Microgrid
Access to base, facility, or equipment is available during outage	Situational: If cyber/IT is not functional, or if microgrid is manually operated, need access[a]
Sufficient battery supply for duration of outage	
Sufficient fuel supply for duration of outage	Situational: Depends on technologies
IT/cyber systems are functional	X
Maintenance does not disrupt power supply	X
Switching mechanism or inverter technologies are functional during outage	X

[a] This condition is considered situational because it must be true only if the microgrid cannot be operated from a remote location.

a dependence on a governing factor for investment or conditions that must be true during an outage for the response option to be considered suitable. "Situational" in a cell indicates that the governing factor applies only under certain conditions. For example, the "Access to base, facility, or equipment is available during an outage" condition must be true if the microgrid cannot be operated from a remote location. Next, a planner can walk through questions pertaining to each applicable governing factor, presented in Figures C.1 and C.2.

The full set of response option governing factors and questions is available in supplemental material that can be requested from the authors.

Figure C.1. Islanded-Mode Enabled Microgrid Investment Questions

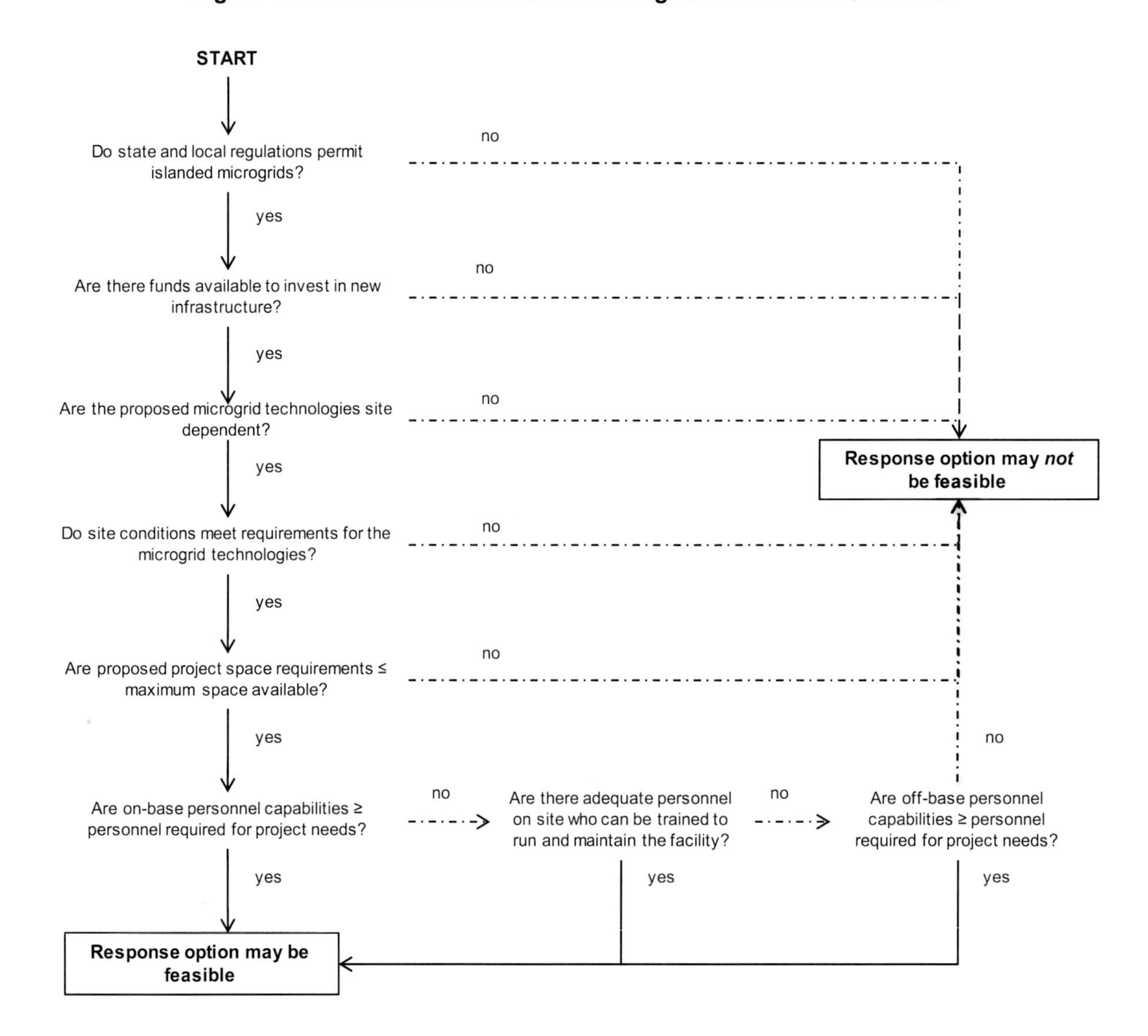

Figure C.2. Islanded-Mode Enabled Microgrid Execution Questions

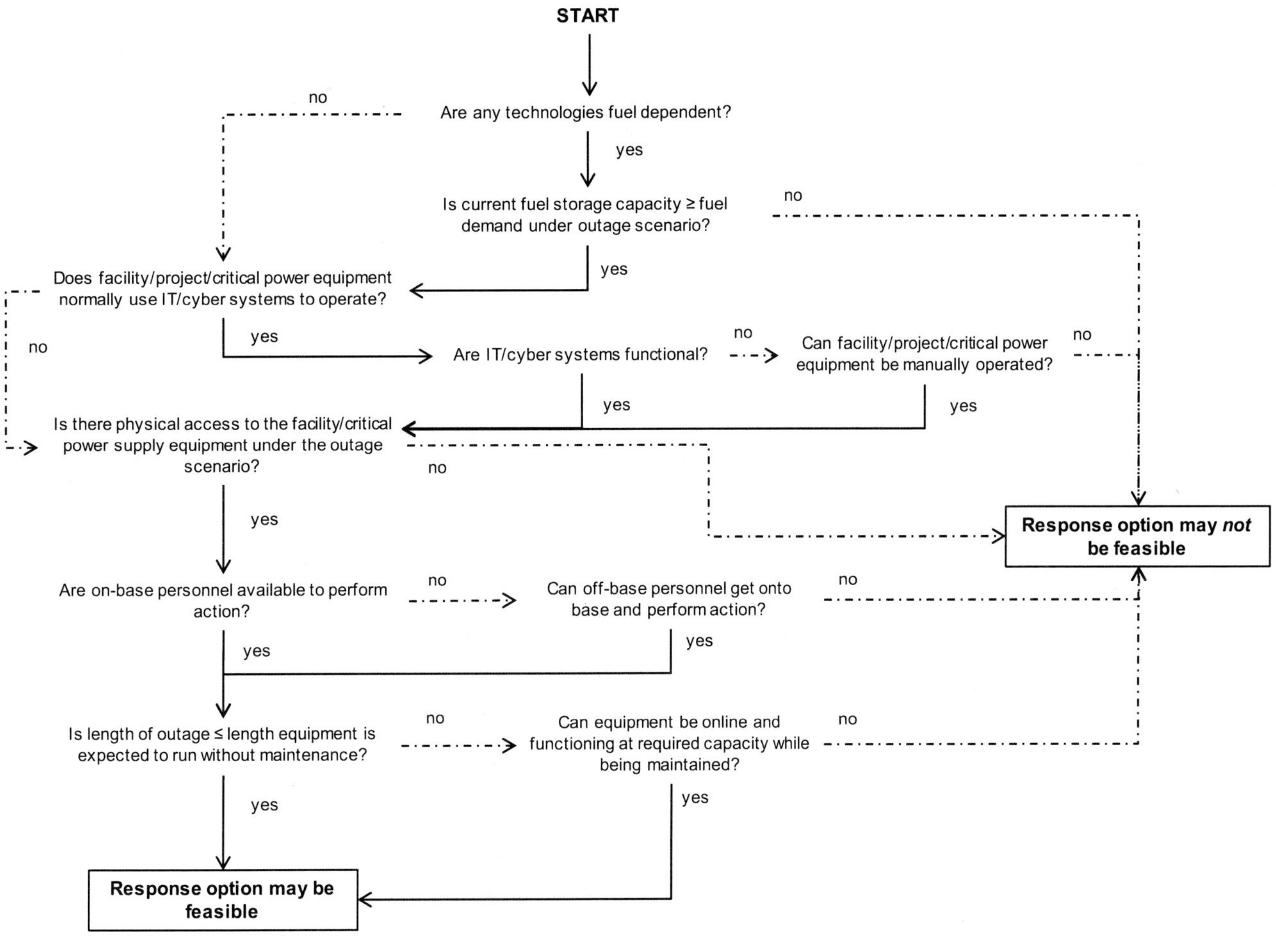

START

Are any technologies fuel dependent?

yes

no

Is current fuel storage capacity ≥ fuel demand under outage scenario?

yes

no

Does facility/project/critical power equipment normally use IT/cyber systems to operate?

yes

no

Are IT/cyber systems functional?

yes

no

Can facility/project/critical power equipment be manually operated?

yes

no

Is there physical access to the facility/critical power supply equipment under the outage scenario?

yes

no

Response option may *not* be feasible

Are on-base personnel available to perform action?

yes

no

Can off-base personnel get onto base and perform action?

yes

no

Is length of outage ≤ length equipment is expected to run without maintenance?

yes

no

Can equipment be online and functioning at required capacity while being maintained?

yes

no

Response option may be feasible

Considering Resources for Response Options

In the preceding sections, we discussed a process for mapping response options to gaps to define a reference set of candidate options and outlined an approach using filters to further assess the suitability of response options to narrow the candidate response set. A helpful next step in evaluating response options is to consider how the response option might be resourced. We present two ways to bin and classify response options: (1) into categories of actions (levers) and (2) by assessing the degree of difficult to implement the response option.

Levers

Different gaps can be addressed through *personnel actions*, *operational changes*, *equipment or technology investment*, or *infrastructure and facility recapitalization.* We refer to these four basic categories of action as *levers*. We bin response options by levers as a means of illuminating different pathways for resourcing response options. For example, costs associated with implementing operational actions will most likely be due to training and the man-hours needed to plan for and implement the response option. It is most likely that these costs will be covered through FY program funds, either by the civil engineering organization or by a mission owner. Table C.7 describes each of these levers and lists associated response options.

Table C.7. Levers for Decisionmakers

Lever	Description	Associated Response Options
Operational changes	Responses that modify the operation of existing systems and equipment	• Perform systematic monitoring • Add cybersecurity controls • Implement demand response, peak demand shaving, or lead shedding • Enable load shifting • Enable moving the mission • Move sensitive equipment out of at-risk locations • Implement preventative maintenance schedules • Make procedural changes • Improve quality of fuel supply contract
Personnel actions	Responses that modify the way existing Air Force or contractor personnel are used or trained; or the addition of manpower authorizations or new contracts	• Add Air Force personnel • Contract out emergency services • Add physical security • Train existing personnel • Enable moving the mission • Add cybersecurity controls
Equipment or technology investment	Responses that involve upgrades to existing equipment or technology; or investment in new equipment or technology	• Automate power distribution for critical load • Add automatic transfer switch for commercial to backup generation • Add prime power generation[a] • Add other on-site power generation[b] • Add battery storage[c] • Add other energy storage[d] • Add diesel power generation • Add EMCS • Increase fuel storage capacity • Add cybersecurity controls • Add PDU • Add islanded-mode enabled microgrid • Add UPS
Infrastructure and facility recapitalization	Responses that involve new, upgrade, or replacement of base infrastructure and building systems	• Upgrade or repair substation • Move electrical lines underground • Add structural hardening • Replace utility line • Add electrical lines • Add islanded-mode enabled microgrid.

[a] Comprises biomass power generation, CHP generation, fuel cells, geothermal power generation, LNG power generation, small nuclear power generation, and methane power generation.
[b] Comprises solar and wind power generation.
[c] Comprises traditional batteries or electric vehicle to grid.
[d] Comprises thermal and chemical energy storage, and flywheels.

Generally, operational changes and personnel actions will be less costly than equipment or technology investments, which are generally less costly than infrastructure and facility recapitalization investments. But this relationship is not absolute and could change for a particular installation, depending on that installation's system architecture.

Assessing Degree of Difficulty: Tiers

Finally, judgments need to be made about level of difficulty. For instance, some responses will be simple and inexpensive to implement. Some will be costly or come against regulatory issues but perhaps well worth the effort required to implement them. To facilitate such judgments, we classify response options into three tiers according to the degree of difficulty of gaining approval, funding, and implementation.

Tier 1 options can generally be implemented in short order at little cost and without onerous approvals or certifications, while implementing Tier 3 options may require substantial investment in time and money, increased personnel requirements, or congressional approval. The most appropriate options, given the energy resilience or reliability challenges the installation is facing, may come from any tier or may be a set of options encompassing multiple tiers. There may also be interim options in lower tiers that can reduce risk while more-comprehensive or cost-effective higher-tier solutions are pursued.

In this work, we have identified a large number of potential options and assigned them to tiers based on existing Air Force policies and processes. These tier assignments are intended as a guide and not meant to be absolute: A solution may be simple to implement at one base but prove challenging at another. Additionally, changing policies and processes can alter the relative difficulty of implementing various solutions.

Tier Definitions

Tier 1

This tier consists of process changes and other actions that are allowable under existing policy and authorities, do not require allocating new funding outside current FY programmed funds, and utilize existing resources.

Tier 1 response options generally fall into two categories: (1) resilience actions identified in existing plans that can be executed during an event with authorities and resources that can be made available within the time frame of the outage (e.g., reassigning personnel or accessing emergency funds) or (2) process changes or other actions to increase reliability or provide more resilience to future events that can be executed under existing authorities and with existing resources. These options require minimal effort to implement or put into contingency plans and can generally be performed at the base level without approval from above. Implementing some of these options during an event, however, may require temporary process changes, formal waivers, or special authorizations.

Tier 2

This tier consists of process changes or other actions that require modest policy adjustments or require financial investment outside current FY programmed funds.

Tier 2 response options require a moderate level of effort to execute. These efforts could include modest policy adjustments, such as changing the prioritization model for project funding or adding manning authorizations. They could require accessing sources of money outside current FY programmed funds, such as the Energy Conservation Investment Program, Federal Energy Management Program, Air Force Energy Initiative, Air Force emergency, O&M, or others. We do not specifically define minimum or maximum thresholds for investment cost in this tier because some programs may be easier to access than others, independent of the amount of money available.

In our review of response options, we used the threshold for O&M-funded unspecified minor construction projects, as stated in U.S. Code Title 10, Section 2805(c), in conjunction with the nature of the approval process to assess classification.[20] For example, if the estimated typical investment of a response option is above the O&M-funded threshold for unspecified minor construction but the approval process is relatively easy, we put the option in Tier 2. Conversely, if the approval process is difficult, we put the option in Tier 3.

Tier 3

This tier consists of process changes or other actions that require substantial investment (recurring and/or nonrecurring) or that involve changes to organizational structure, authorities, or the mechanisms of funding and resource allocation that would require issuing new policy or making major changes to existing policy.

Tier 3 response options require a high level of effort to execute for one or more of the following reasons: the nature of the approval process (e.g., congressional approval for military construction [MILCON] projects), the high dollar investment required (both defense budget funding streams and third-party financing), Air Force contractual obligations (e.g., utility privatization and enhanced use lease arrangements), or the number of stakeholders or organizations required to enact the policy change or to program and allocate funding. Making major changes to existing policy or enacting new policy may be structural or systemic. For example, new policy that imposes strict greenhouse gas emission standards may require systemic operational changes at the base level without changing existing structure, authorities, or funding allocation mechanisms.

Tier Classification and Justifications

Table C.8 lays out and justifies our tier assignments for each of the response options we examined. These assignments are intended as a guide and not meant to be absolute: A solution

---

[20] As of the writing of this document, the threshold was $1,000,000.

may be simple to implement at one base but challenging at another. Additionally, changing policies and processes can alter the relative difficulty of implementing various solutions.

Table C.8. Tier Classification and Justification

Option	Tier 1	Tier 2	Tier 3	Tier Justification
Perform systematic monitoring	X			Most likely can be implemented easily
Implement demand response, peak demand shaving, or load shedding	X			Most likely can be implemented easily
Implement preventative maintenance schedules	X			Most likely can be implemented within a short time frame and without approval from authorities above the base level
Make procedural changes	X			Most likely can be implemented within a short time frame and without approval from authorities above the base level
Train existing personnel	X			Most likely minimal time and dollar investment
Add Air Force personnel	X	X		Tier 1 if calling in off-duty personnel or borrowing billets from other organizations during an outage; Tier 2 if adding manning authorizations
Add cybersecurity controls	X	X		Tier 1 if implementing simple procedural changes, such as access control removable drive scanning; Tier 2 if need to create new position, invest in equipment or technology and coordinate with other organizations (e.g., communications group or security) to implement program
Increase liquid fuel storage capacity	X	X		Tier 1 if adding temporary storage; Tier 2 if adding permanent storage capacity
Add physical security	X	X		Tier 1 if have enough people to add to security rounds; Tier 2 if need to add manning authorizations
Automate power distribution for critical load		X		Most likely requires additional funding outside FY programmed funds
Add automatic transfer switch for commercial to backup generation		X		Most likely requires additional funding outside FY programmed funds
Add batteries		X		Most likely requires additional funding outside FY programmed funds
Contract out emergency services		X		Requires contracting mechanism
Add diesel power generation		X		Most likely requires additional funding outside FY programmed funds
Add EMCS		X		Most likely requires additional funding outside FY programmed funds
Enable load shifting		X		Most likely requires additional funding outside FY programmed funds
Enable moving the mission		X		Handoff process can be complicated; may require additional manning authorizations or equipment.
Move critical power equipment out of at-risk locations		X		Most likely requires additional funding outside FY programmed funds

Table C.8—Continued

Option	Tier 1	Tier 2	Tier 3	Tier Justification
Add PDU		X		Most likely requires additional funding outside FY programmed funds
Add UPS		X		Most likely requires additional funding outside FY programmed funds
Add electric vehicle to grid		X	X	Depends on project costs
Upgrade or repair substation		X	X	Depends on project costs, small repairs might be Tier 2, while upgrades might be considered Tier 3 if MILCON project
Improve quality of fuel supply contract		X	X	Need to involve the Defense Logistics Agency; may require changes to enterprisewide contract
Add solar photovoltaic power generation		X	X	Depends on system size, project costs, and funding source
Add biomass power generation plant			X	Most likely considered MILCON project
Add CHP generation plant			X	Most likely considered MILCON project
Add energy storage (other than batteries)			X	Most likely considered MILCON project
Add fuel cell			X	Most likely considered MILCON project
Add geothermal power generation			X	Most likely considered MILCON project
Add islanded-mode enabled microgrid			X	Most likely considered MILCON project
Add LNG power generation plant			X	Most likely considered MILCON project
Add methane power generation plant			X	Most likely considered MILCON project
Move electrical lines underground			X	Most likely considered MILCON project
Add structural hardening			X	Most likely considered MILCON project
Add small nuclear power generation plant			X	Most likely considered MILCON project; permitting and approval process very difficult
Upgrade or replace utility lines			X	Most likely considered MILCON project
Add wind power generation			X	Most likely high dollar investment; permitting and approval process can be lengthy
Add redundant electrical lines			X	Most likely considered MILCON project

References

AFI—*See* Air Force Instruction.

Air Force Instruction 10-208, *Continuity of Operations (COOP) Program*, October 16, 2013. As of July 18, 2017: http://static.e-publishing.af.mil/production/1/afmc/publication/afmci10-208/afmci10-208.pdf

Air Force Instruction 23-201, *Fuels Management*, June 20, 2014. As of September 23, 2016: http://static.e-publishing.af.mil/production/1/af_a4_7/publication/afi23-201/afi23-201.pdf

Air Force Instruction 32-1023, *Designing and Constructing Military Construction Projects*, November 19, 2015. As of July 17, 2017: http://static.e-publishing.af.mil/production/1/af_a4/publication/afi32-1023/afi32-1023.pdf

Air Force Instruction 32-1062, *Electrical Systems, Power Plants and Generators*, January 15, 2015. As of September 23, 2016: http://static.e-publishing.af.mil/production/1/af_a4/publication/afi32-1062/afi32-1062.pdf

Air Force Instruction 32-1063, *Electric Power Systems*, June 10, 2005 (superseded by Air Force Instruction 32-1062).

Air Force Instruction 32-9005, *Real Property Accountability and Reporting,* March 4, 2015. As of September 5, 2017: http://static.e-publishing.af.mil/production/1/af_a4/publication/afi32-9005/afi32-9005.pdf

Air Force Instruction 32-10142, *Facilities Board*, May 14, 2013. As of September 23, 2016: http://static.e-publishing.af.mil/production/1/af_a4/publication/afi32-10142/afi32-10142.pdf

Air Force Pamphlet 32-10144_AFGM2016-01, "Implementing Utilities at U.S. Air Force Installations," with memorandum, September 16, 2016. As of February 15, 2017: http://static.e-publishing.af.mil/production/1/af_a4/publication/afpam32-10144/afpam32-10144.pdf

Air Force Space Command, "Resiliency and Disaggregated Space Architectures," white paper, 2013. As of September 26, 2016: https://fas.org/spp/military/resiliency.pdf

Allenby, Brad, and Jonathan Fink, "Toward Inherently Secure and Resilient Societies," *Science*, Vol. 309, No. 5737, August 2005, pp. 1034–1035.

Anderies, John M., Carl Folke, Brian Walker, and Elinor Ostrom, "Aligning Key Concepts for Global Change Policy: Robustness, Resilience, and Sustainability," *Ecology and Society*,

Vol. 18, No. 2, 2013. As of September 22, 2016: http://www.ecologyandsociety.org/vol18/iss2/art8/

Bernardo, Jose, Bryan Boling, Philippe A. Bonnefoy, Graham Burdette, R. John Hansman, Michelle Kirby, Dongwook Lim, Dimitri Mavris, Aleksandra Mozdzanowska, Taewoo Nam, Holger Pfaender, Ian A. Waitz, and Brian Yutko, "CO_2 Emission Metrics for Commercial Aircraft Certification: A National Airspace System Perspective," Cambridge, Mass.: Partnership for AIR Transportation Noise and Emissions Reduction, Massachusetts Institute of Technology, PARTNER-COE-2012-002, March 2012. As of September 26, 2016: http://partner.mit.edu/projects/metrics-aviation-co2-standard

Bollen, Mathias H. J., "What Is Power Quality?" *Electric Power Systems Research*, Vol. 66, No. 1, July 2003, pp. 5–14.

Bompard, E., R. Napoli, and F. Xue, "Extended Topological Approach for the Assessment of Structural Vulnerability in Transmission Networks," *Generation, Transmission & Distribution*, Vol. 4, No. 6, May 2010, pp. 716–724.

Bonikowsky, Laura Neilson, and Niko Block, "Ice Storm of 1998," *Historica Canada*, February 11, 2016. As of September 26, 2016: http://www.thecanadianencyclopedia.ca/en/article/ice-storm-1998/

Chermack, Thomas J., *Scenario Planning in Organizations: How to Create, Use, and Assess Scenarios*, San Francisco: Berrett-Koehler Publishers, 2011.

Chermack, Thomas J., Susan A. Lynham, and Wendy E. A. Ruona, "A Review of Scenario Planning Literature," *Futures Research Quarterly*, Vol. 17, No. 2, Summer 2001, pp. 7–31.

Committee on Increasing National Resilience to Hazards and Disasters, Committee on Science, Engineering, and Public Policy, and The National Academies, *Disaster Resilience: A National Imperative*, Washington, D.C.: The National Academies Press, 2012. As of September 22, 2016: https://www.nap.edu/catalog/13457/disaster-resilience-a-national-imperative

Davis, Paul K., *Lessons from RAND's Work on Planning Under Uncertainty for National Security*, Santa Monica, Calif.: RAND Corporation, TR-1249-OSD, 2012. As of September 26, 2016: http://www.rand.org/pubs/technical_reports/TR1249.html

Delta Stewardship Council, "Delta Levees Investment Strategy Decision Support Tool," website, Public Version 2.2_beta, February 2017. As of July 18, 2017: http://deltacouncil.ca.gov/dlis-decision-support-tool

Department of the Air Force, "Engineering Technical Letter (ETL) 13-4 (Change 1): Standby Generator Design, Maintenance, and Testing Criteria," May 15, 2014. As of September 5,

2017:
https://www.wbdg.org/FFC/AF/AFETL/etl_13_4.pdf

Department of Defense Directive 3020.40, *DoD Policy and Responsibilities for Critical Infrastructure*, January 14, 2010a. As of September 5, 2017:
http://policy.defense.gov/Portals/11/Documents/hdasa/newsletters/302040p.pdf

Dewar, James A., *Assumption-Based Planning: A Tool for Reducing Avoidable Surprises*, Cambridge, United Kingdom: Cambridge University Press, 2002.

Dewar, James A., Carl H. Builder, William M. Hix, and Morlie Levin, *Assumption-Based Planning: A Planning Tool for Very Uncertain Times*, Santa Monica, Calif.: RAND Corporation, MR-114-A, 1993. As of September 26, 2016:
http://www.rand.org/pubs/monograph_reports/MR114.html

DoD—*See* U.S. Department of Defense.

Donley, Michael B., "Air Force Policy Memorandum to AFPD 10-24, Air Force Critical Infrastructure Program (CIP), 28 April 2006," Washington, D.C., January 6, 2012. As of September 26, 2016:
http://static.e-publishing.af.mil/production/1/af_a3_5/publication/afpd10-24/afpd10-24.pdf

Environment and Climate Change Canada, "Canada's Top Ten Weather Stories of 1998," May 18, 2013. As of September 26, 2016:
http://www.ec.gc.ca/meteo-weather/default.asp?lang=En&n=3DED7A35-1

Greene, Wedge, and Barbara Lancaster, "Carrier-Grade: Five Nines, the Myth and the Reality," *Pipeline*, Vol. 3, No. 11, 2006. As of September 27, 2016:
http://www.pipelinepub.com/0407/pdf/Article%204_Carrier%20Grade_LTC.pdf

Gunderson, Lance H., "Ecological Resilience—In Theory and Application," *Annual Review of Ecology and Systematics*, Vol. 31, 2000, pp. 425–439.

Haimes, Yacov Y., "On the Definition of Resilience in Systems," *Risk Analysis*, Vol. 29, No. 4, 2009, pp. 498–501.

Hirst, Eric, and Brendan Kirby, "Bulk-Power Basics: Reliability and Commerce," Oak Ridge, Tenn.: Consulting in Electric-Industry Restructuring, January 24, 2000. As of September 26, 2016:
http://www.consultkirby.com/files/RAPBPBasics.pdf

Holland, Jeffery P., "Engineered Resilient Systems (ERS) Overview," briefing, December 2013. As of September 27, 2016:
http://defenseinnovationmarketplace.dtic.mil/resources/ERS_Overview_2DEC2013-Final.pdf

Holling, C. S., "Resilience and Stability of Ecological Systems," *Annual Review of Ecology and Systematics*, Vol. 4, 1973, pp. 1–23.

Ibanez, Eduardo, Steven Lavrenz, Konstantina Gkritza, Diego A. Mejia-Giraldo, Venkat Krishnan, and James D. McCalley, "Resilience and Robustness in Long-Term Planning of the National Energy and Transportation Systems," *International Journal of Critical Infrastructures*, Vol. 12, No. 1, 2016, pp. 82–103.

Industrial Control Systems Cyber Emergency Response Team, "Cyber-Attack Against Ukrainian Critical Infrastructure," alert, February 25, 2016. As of September 23, 2016: https://ics-cert.us-cert.gov/alerts/IR-ALERT-H-16-056-01

Intergovernmental Panel on Climate Change, "Summary for Policymakers," in Christopher B. Field et al., eds., *Climate Change 2014: Impacts, Adaptation and Vulnerability*, New York: Cambridge University Press, 2014. As of September 26, 2016: http://www.ipcc.ch/pdf/assessment-report/ar5/wg2/ar5_wgII_spm_en.pdf

International Electrotechnical Commission, IEC 61000-3-3, "Limits—Limitation of Voltage Changes, Voltage Fluctuations and Flicker in Public Low-Voltage Supply Systems, for Equipment with Rated Current ≤16 A per Phase and Not Subject to Conditional Connection," Geneva, Switzerland, 2008.

IPCC—*See* Intergovernmental Panel on Climate Change.

Jennings, Barbara J., Eric D. Vugrin, and Deborah K. Belasich, "Resilience Certification for Commercial Buildings: A Study of Stakeholder Perspectives," *Environment Systems and Decisions*, Vol. 33, No. 2, June 2013, pp. 184–194.

Judson, N., A. L. Pina, E. V. Dydek, S. B. Van Broekhoven, A. S. Castillo, *Application of a Resilience Framework to Military Installations: A Methodology for Energy Resilience Business Case Decisions*, Lexington, Mass.: Lincoln Laboratory, Massachusetts Institute of Technology, Technical Report 1216, October 4, 2016.

Keogh, Miles, and Christina Cody, "Resilience in Regulated Utilities," Washington, D.C.: National Association of Regulatory Utility Commissioners, November 2013. As of September 26, 2016: https://pubs.naruc.org/pub/536F07E4-2354-D714-5153-7A80198A436D

Kueck, John D., Brendan Kirby, Philip Overholt, and Lawrence Markel, "Measurement Practices for Reliability and Power Quality," Oak Ridge, Tenn.: Oak Ridge National Laboratory, June 2004.

Lempert, Robert J., Steven W. Popper, and Steven C. Bankes, *Shaping the Next One Hundred Years: New Methods for Quantitative, Long-Term Policy Analysis*, Santa Monica, Calif.:

RAND Corporation, MR-1626-RPC, 2003. As of September 26, 2016: http://www.rand.org/pubs/monograph_reports/MR1626.html

Lempert, Robert J., Steven W. Popper, David G. Groves, Nidhi Kalra, Jordan R. Fischbach, Steven C. Bankes, Benjamin P. Bryant, Myles T. Collins, Klaus Keller, Andrew Hackbarth, Lloyd Dixon, Tom LaTourrette, Robert T. Reville, Jim W. Hall, Christophe Mijere, and David J. McInerney, "Making Good Decisions Without Predictions: Robust Decision Making for Planning Under Deep Uncertainty," Santa Monica, Calif.: RAND Corporation, RB-9701, 2013. As of September 26, 2016: http://www.rand.org/pubs/research_briefs/RB9701.html

Martínez-Anido, C. Brancucci, R. Bolado, L. De Vries, G. Fulli, M. Vandenbergh, and M. Masera, "European Power Grid Reliability Indicators, What Do They Really Tell?" *Electric Power Systems Research*, Vol. 90, May 2012, pp. 79–84.

Masten, Ann S., "Ordinary Magic: Lessons from Research on Resilience in Human Development," *Education Canada*, Vol. 49, No. 3, Summer 2009, pp. 28–32.

McCarthy, Ryan W., Joan M. Ogden, and Daniel Sperling, "Assessing Reliability in Energy Supply Systems," *Energy Policy*, Vol. 35, No. 4, April 2007, pp. 2151–2162.

McDonald, James R., and Kishor C. Mehta, "A Recommendation for an Enhanced Fujita Scale (EF-Scale)," Lubbock, Tex.: Wind Science and Engineering Center, Texas Tech University, 2006.

McNerney, Michael J., Jefferson P. Marquis, S. Rebecca Zimmerman, and Ariel Klein, *SMART Security Cooperation Objectives: Improving DoD Planning and Guidance*, Santa Monica, Calif.: RAND Corporation, RR-1430-OSD, 2016. As of September 28, 2016: http://www.rand.org/pubs/research_reports/RR1430.html

Mooallem, Jon, "Squirrel Power!" *New York Times*, August 31, 2013. As of September 26, 2016: http://www.nytimes.com/2013/09/01/opinion/sunday/squirrel-power.html

National Defence and the Canadian Armed Forces, "Operation Recuperation," news release, January 12, 1998. As of September 23, 2016: http://www.forces.gc.ca/en/news/article.page?doc=operation-recuperation-news-release/hnlhlxio

North American Electric Reliability Corporation, *Reliability Assessment Guidebook*, Vers. 3.1, Atlanta, Ga., August 2012. As of September 18, 2017: http://www.nerc.com/files/Reliability%20Assessment%20Guidebook%203%201%20Final.pdf

North American Electric Reliability Corporation, "Adequate Level of Reliability," May 2013. As
North American Electric Reliability Corporation, "Adequate Level of Reliability," May

2013. As of September 22, 2016: http://www.nerc.com/pa/Stand/Resources/Documents/Adequate_Level_of_Reliability_Definition_(Informational_Filing).pdf

North American Electric Reliability Organization, *2015 Long-Term Reliability Assessment*, Atlanta, Ga., December 2015. As of September 27, 2016: http://www.nerc.com/pa/RAPA/ra/Reliability%20Assessments%20DL/2015LTRA%20-%20Final%20Report.pdf

Osborn, Julie, and Cornelia Kawann, "Reliability of the U.S. Electricity System: Recent Trends and Current Issues," Energy Analysis Department, Ernest Orlando Lawrence Berkeley National Laboratory, Berkeley, Calif., LBNL-47043, 2001.

Pillay, P., and M. Manyage, "Definitions of Voltage Unbalance," *IEEE Power Engineering Review*, May 2001. As of September 26, 2016: http://users.encs.concordia.ca/~pillay/16.pdf

Pinelli, Jean-Paul, Emil Simiu, Kurt Gurley, Chelakara Subramanian, Liang Zhang, Anne Cope, James J. Filliben, and Shahid Hamid, "Hurricane Damage Prediction Model for Residential Structures," *Journal of Structural Engineering*, Vol. 130, No. 11, November 2004, pp. 1685–1691.

Presidential Policy Directive 21, Critical Infrastructure Security and Resilience, February 12, 2013. As of September 22, 2016: https://obamawhitehouse.archives.gov/the-press-office/2013/02/12/presidential-policy-directive-critical-infrastructure-security-and-resil

Public Law 112-81, National Defense Authorization Act for Fiscal Year 2012, December 31, 2011. As of September 22, 2016: https://www.congress.gov/112/plaws/publ81/PLAW-112publ81.pdf

Radakovich, Amy L., Alex J. Ferguson, and John Boatwright, "Field Survey of Earthquake Effects from the Magnitude 4.0 Southern Maine Earthquake of October 16, 2012," Reston, Va.: U.S. Geological Survey, Open-File Report 2016-1071, 2016. As of September 26, 2016: http://pubs.usgs.gov/of/2016/1071/ofr20161071.pdf

RAND Corporation, "Adapting to a Changing Colorado River," webpage, undated a. As of July 18, 2017: http://www.rand.org/jie/infrastructure-resilience-environment/projects/colorado-river-basin/interactive-brief.html

RAND Corporation, "Robust Decision Making," webpage, undated b. As of July 18, 2017: http://www.rand.org/topics/robust-decision-making.html

Roe, Emery, and Paul Schulman, "Toward a Comparative Framework for Measuring Resilience in Critical Infrastructure Systems," *Journal of Comparative Policy Analysis: Research and Practice*, Vol. 14, No. 2, April 2012, pp. 114–125.

Rouse, Greg, and John Kelly, "Electricity Reliability: Problems, Progress and Policy Solutions," Galvin Electricity Initiative, February 2011. As of September 26, 2016: http://www.galvinpower.org/sites/default/files/Electricity_Reliability_031611.pdf

Rutter, Michael, "Developing Concepts in Developmental Psychopathology," in James J. Hudziak, ed., *Developmental Psychopathology and Wellness: Genetic and Environmental Influences*, Washington, D.C.: American Psychiatric Publishing, 2008, pp. 3–22.

Samaras, Constantine, and Henry H. Willis, *Capabilities-Based Planning for Energy Security at Department of Defense Installations*, Santa Monica, Calif.: RAND Corporation, RR-162-RC, 2013. As of March 3, 2016: http://www.rand.org/pubs/research_reports/RR162.html

Sandia National Laboratories, "Motivating Business to Design a More Resilient Nation, One Building at a Time," new release, July 23, 2013. As of September 22, 2016: https://share.sandia.gov/news/resources/news_releases/motivating-business-to-design-a-more-resilient-nation-one-building-at-a-time/

Sandia National Laboratories, "Resilience Metrics for Energy Transmission and Distribution Infrastructure," briefing, Quadrennial Energy Review Workshop, June 10, 2014. As of September 23, 2016: http://energy.gov/sites/prod/files/2015/01/f19/QER%20Workshop%20June%2010%202014%20Posted.pdf

Savitz, Scott, Henry H. Willis, Aaron Davenport, Martina Melliand, William Sasser, Elizabeth Tencza, and Dulani Woods, *Enhancing U.S. Coast Guard Metrics*, Santa Monica, Calif.: RAND Corporation, RR-1173-USCG, 2015. As of September 26, 2016: http://www.rand.org/pubs/research_reports/RR1173.html

Schwartz, Peter, *The Art of the Long View: Planning for the Future in an Uncertain World*, New York: Currency, 1996.

SEDAPAL, "A Robust Strategy for Implementing Lima's Long-Term Water Resources Master Plan," tool, May 11, 2015. As of July 18, 2017: https://public.tableau.com/profile/david.groves1600#!/vizhome/SEDAPAL_PDT-2015_05_10_0/SEDAPALPDT

Sutcliffe, Kathleen M., and Timothy J. Vogus, "Organizing for Resilience," in Kim S. Cameron, Jane E. Dutton, and Robert E. Quinn, eds., *Positive Organizational Scholarship: Foundations of a New Discipline*, San Francisco: Berrett-Koehler Publishers, 2003, pp. 94–110.

Teodorescu, Remus, and Marco Liserre, *Grid Converters for Photovoltaic and Wind Power Systems*, Chichester, United Kingdom: John Wiley & Sons, 2011.

Trivedi, Kishor, Dong Seong Kim, and Rahul Ghosh, "Resilience in Computer Systems and Networks," *2009 IEEE/ACM International Conference on Computer-Aided Design Digest of Technical Papers*, November 2009, pp. 74–76.

University of Kansas Information and Telecommunication Technology Center, "ResiliNets Architecture Definitions," webpage, 2014. As of September 22, 2016: https://wiki.ittc.ku.edu/resilinets_wiki/index.php/Definitions

U.S. Air Force, "AF Tests First All-Electric Vehicle Fleet in California," news release, November 14, 2014. As of July 31, 2016: http://www.af.mil/News/ArticleDisplay/tabid/223/Article/554343/af-tests-first-all-electric-vehicle-fleet-in-california.aspx

U.S. Code Title 10, Section 2805(c), Unspecified Minor Construction.

U.S. Department of Defense, *Quadrennial Defense Review Report*, February 2010. As of September 22, 2016: http://www.defense.gov/Portals/1/features/defenseReviews/QDR/QDR_as_of_29JAN10_1600.pdf

U.S. Department of Defense, *Mission Assurance Strategy*, April 2012a. As of September 22, 2016: http://policy.defense.gov/Portals/11/Documents/MA_Strategy_Final_7May12.pdf

U.S. Department of Defense, "Pentagon Officials Provide Storm Response Update," news article, November 7, 2012b. As of September 23, 2016: http://archive.defense.gov/news/newsarticle.aspx?id=118473

U.S. Department of Defense, "Energy Department Loans Fuel to DOD for Connecticut Distribution," news article, November 10, 2012c. As of September 23, 2016: http://archive.defense.gov/news/newsarticle.aspx?id=118501

U.S. Department of Energy, "Hurricane Sandy-Nor'easter Situation Report #13," December 3, 2012. As of September 23, 2016: http://www.oe.netl.doe.gov/docs/SitRep13_Sandy-Nor'easter_120312_300PM.pdf

U.S. Department of Energy, *Quadrennial Energy Review: Energy Transmission, Storage, and Distribution Infrastructure*, 2015. As of September 5, 2017: https://energy.gov/sites/prod/files/2015/07/f24/QER%20Full%20Report_TS%26D%20April%202015_0.pdf

van der Heijden, Kees, "Scenarios, Strategies and the Strategy Process," Breukelen, The Netherlands: Nijenrode University Centre for Organisational Learning and Change,

No. 1997-01, January 1997. As of September 26, 2016:
http://citeseerx.ist.psu.edu/viewdoc/download?doi=10.1.1.202.9185&rep=rep1&type=pdf

Vijayaraghavan, G., Mark Brown, and Malcolm Barnes, *Practical Grounding, Bonding, Shielding and Surge Protection*, Waltham, Mass.: Butterworth-Heinemann, 2004.

Voorspools, Kris R., and William D. D'Haeseleer, "Reliability of Power Stations: Stochastic Versus Derated Power Approach," *International Journal of Energy Research*, Vol. 28, No. 2, February 2004, pp. 117–128.

Walker, Brian, C. S. Holling, Stephen R. Carpenter, and Ann Kinzig, "Resilience, Adaptability and Transformability in Social-Ecological Systems," *Ecology and Society*, Vol. 9, No. 2, 2004. As of September 22, 2016:
http://www.ecologyandsociety.org/vol9/iss2/art5/

Watson, Jean-Paul, Ross Guttromson, Cesar Silva-Monroy, Robert Jeffers, Katherine Jones, James Ellison, Charles Rath, Jared Gearhart, Dean Jones, Tom Corbet, Charles Hanley, and La Tonya Walker, *Conceptual Framework for Developing Resilience Metrics for the Electricity, Oil, and Gas Sectors in the United States*, Albuquerque, N.M.: Sandia National Laboratories, September 2014. As of September 22, 2016:
http://energy.gov/sites/prod/files/2015/02/f20/EnergyResilienceRpt-Sandia-Sep2014.pdf

Wei, Dong, and Kun Ji, "Resilient Industrial Control System (RICS): Concepts, Formulation, Metrics, and Insights," in Institute of Electrical and Electronics Engineers, *3rd International Symposium on Resilient Control Systems (ISRCS)*, 2010, pp. 15–22.

Wieland, Andreas, and Carl Marcus Wallenburg, "The Influence of Relational Competencies on Supply Chain Resilience: A Relational View," *International Journal of Physical Distribution and Logistics Management*, Vol. 43, No. 4, 2013, pp. 300–320.

Willis, Henry H., and Kathleen Loa, *Measuring the Resilience of Energy Distribution Systems*, Santa Monica, Calif.: RAND Corporation, RR-883-DOE, 2015. As of September 20, 2016:
http://www.rand.org/pubs/research_reports/RR883.html

Yeddanapudi, Sree Rama Kumar, "Distribution System Reliability Evaluation," briefing slides, July 2012. As of September 26, 2016:
http://www.slideserve.com/oshin/distribution-system-reliability-evaluation

Young, Stephanie, Henry H. Willis, Melinda Moore, and Jeffrey Engstrom, *Measuring Cooperative Biological Engagement Program (CBEP) Performance: Capacities, Capabilities, and Sustainability Enablers for Biorisk Management and Biosurveillance*, Santa Monica, Calif.: RAND Corporation, RR-660-OSD, 2014. As of September 26, 2016:
http://www.rand.org/pubs/research_reports/RR660.html

Zetter, Kim, "Inside the Cunning, Unprecedented Hack of Ukraine's Power Grid," *Wired*, March 3, 2016. As of September 23, 2016: https://www.wired.com/2016/03/inside-cunning-unprecedented-hack-ukraines-power-grid/